1st Workshop on Mathematical Natural Language Processing (MathNLP 2022)

AA004827

Abu Dhabi, United Arab Emirates and Online
8 December 2022

ISBN: 978-1-7138-9113-0

Printed from e-media with permission by:

Curran Associates, Inc.
57 Morehouse Lane
Red Hook, NY 12571

CURRAN ASSOCIATES INC.
proceedings
.com

Some format issues inherent in the e-media version may also appear in this print version.

Copyright© (2022) by the Association for Computational Linguistics
All rights reserved.

Printed with permission by Curran Associates, Inc. (2024)

For permission requests, please contact the Association for Computational Linguistics
at the address below.

Association for Computational Linguistics
209 N. Eighth Street
Stroudsburg, Pennsylvania 18360

Phone: 1-570-476-8006
Fax: 1-570-476-0860

acl@aclweb.org

Additional copies of this publication are available from:

Curran Associates, Inc.
57 Morehouse Lane
Red Hook, NY 12571 USA
Phone: 845-758-0400
Fax: 845-758-2633
Email: curran@proceedings.com
Web: www.proceedings.com

MathNLP 2022

1st Workshop on Mathematical Natural Language Processing

Proceedings of the Workshop

December 8, 2022

Introduction

Articulating mathematical arguments is a fundamental part of scientific reasoning and communication. Across many disciplines, expressing relations and interdependencies between quantities (usually in an equational form)is at the center of scientific argumentation. One can easily find examples of mathematical discourse across different scientific contributions and textbooks. Nevertheless, the application of contemporary models for performing inference over mathematical text still needs to be explored despite its importance.

Creating methods and models that can understand mathematical text and discourse will pave the path toward developing systems capable of complex mathematical inference, leading to automated scientific discovery in fields that depend on mathematical knowledge. However, there are still technical gaps that need to be addressed, such as the availability of datasets and evaluation tasks, techniques for the joint interpretation of different modalities present in the mathematical text (equational and natural language), the understanding of unique aspects of mathematical discourse and multi-hop models for mathematical inference. The Workshop on Mathematical Natural Language Processing (MathNLP) represents a community-building venue for addressing these challenges by connecting experts at the intersection of Mathematics and Natural Language Processing.

The first edition of MathNLP received a total of 12 submissions, accepting 7 of them for publication in the proceedings. In addition, MathNLP welcomed a total of 4 oral presentations from papers accepted to appear in Findings of EMNLP 2022.

We would like to thank our keynote speakers for their contribution to the program and the members of the program committee for their valuable and high-quality reviews. All submissions have benefited from their expert feedback. Their timely contribution was the basis for accepting an excellent list of papers and making the first edition of MathNLP a success.

Deborah Ferreira, Marco Valentino, Andre Freitas, Sean Welleck, Moritz Schubotz.

1st MathNLP Organizers

November 2022

Organizing Committee

Organizers

Deborah Ferreira, The MathWorks
Marco Valentino, Idiap Research Institute
Andre Freitas, University of Manchester and Idiap Research Institute
Sean Welleck, University of Washington and Allen Institute for Artificial Intelligence
Moritz Schubotz, University of Wuppertal

Program Committee

Reviewers

Somak Aditya, Akiko Aizawa, Takuto Asakura

Matthew Finlayson

Jordan Meadows

Mokanarangan Thayaparan

Keynote Talk: LLMs-as-a-Service: Harnessing the power of Foundation Models for Challenging Reasoning Problems

Ashwin Kalyan Vijayakumar
Allen Institute for Artificial Intelligence

Abstract: AI systems built on top of foundation models achieve state-of-the-art performance on a wide range of tasks making them the one of the most versatile and dependable AI technology. However, even for these systems, hard reasoning problems — ones that require mathematical and algorithmic reasoning in addition to more general skills like language understanding, commonsense reasoning and computer vision — pose a significant challenge. First, I will discuss the successes and limitations of state-of-the-art LLMs on hard reasoning problems like fermi problems and challenging math word problems — encouraging the broader AI community to address this challenge in AI reasoning. Next, I propose "LLMs-as-a-Service", a compositional and neuro-symbolic strategy to develop the next generation of AI solutions that achieve best-of-both-worlds — harness the capacity of powerful foundational models while at the same time overcoming their shortcomings in producing well-reasoned, consistent answers.

Bio: Ashwin Kalyan is a scientist connecting AI, innovation and research. He led and contributed to research projects and technologies that have resulted in new perspectives of integrating AI systems with practice (e.g. neuro-symbolic approaches for program synthesis, novel decoding strategies for language models) that have impacted industry practices in addition to the wider research community. Currently, he is a researcher at the Allen Institute of Artificial Intelligence where he investigates the abilities and limitations of foundation models, especially in the context of hard reasoning problems that require mathematical and algorithmic reasoning. He has authored 20+ publications in top-tier AI conferences (e.g. NeurIPS, ICML, CVPR, ACL, EMNLP) and was recognized by the prestigious JP Morgan PhD Fellowship. He obtained his PhD from Georgia Institute of Technology and prior to that, B.Tech from National Institute of Technology Karnataka. He started the "student researcher" program at AI2, a research apprenticeship initiative that nurtures scientific talent by providing aspiring researchers (including undergraduate and PhD students) a peek into cutting-edge AI research. He serves as the technical advisor for Youth for Creativity and Excellence (YCEF), a privately funded non-profit organization that promotes scientific, cultural and creative pursuits in India.

Table of Contents

Program

Thursday, December 8, 2022

09:00 - 09:15 *Opening Remarks*

09:15 - 10:30 *Oral Session 1*

Tracing and Manipulating intermediate values in Neural Math Problem Solvers
Yuta Matsumoto, Benjamin Heinzerling, Masashi Yoshikawa and Kentaro Inui

Evaluating Token-Level and Passage-Level Dense Retrieval Models for Math Information Retrieval
Jimmy Lin, YUQING XIE, Jheng-Hong Yang and Wei Zhong

Investigating Math Word Problems using Pretrained Multilingual Language Models
Minghuan Tan, Lei Wang, Lingxiao Jiang and Jing Jiang

Induced Natural Language Rationales and Interleaved Markup Tokens Enable Extrapolation in Large Language Models
Mirelle Candida Bueno, Carlos Gemmell, Jeff Dalton, Roberto Lotufo and Rodrigo Nogueira

Towards Autoformalization of Mathematics and Code Correctness: Experiments with Elementary Proofs
Garett Cunningham, Razvan Bunescu and David Juedes

10:30 - 11:00 *Coffee Break 1*

11:00 - 11:45 *Invited Talk 1: Ashwin Kalyan: LLMs-as-a-Service: Harnessing the power of Foundation Models for Challenging Reasoning Problems*

11:45 - 12:30 *Oral Session 2*

Textual Enhanced Contrastive Learning for Solving Math Word Problems
Sadao Kurohashi, Fei Cheng, Zhuoyuan Mao, Qianying Liu and Yibin Shen

Multi-View Reasoning: Consistent Contrastive Learning for Math Word Problem
Weiming Lu, Qingpeng Nong, Zeqi Tan, Xiaoxia Cheng, Yanna Ma, Yongliang Shen and Wenqi Zhang

LogicSolver: Towards Interpretable Math Word Problem Solving with Logical Prompt-enhanced Learning
Xiaodan Liang, Liang Lin, Jiaqi Chen, Jinghui Qin and Zhicheng Yang

Thursday, December 8, 2022 (continued)

12:30 - 14:00 *Lunch*

14:00 - 14:45 *Building the Automated Mathematician: an NLP Perspective*

14:45 - 15:30 *Oral Session 3*

Numerical Correlation in Text
Daniel Spokoyny, Chien-Sheng Wu and Caiming Xiong

Extracting Operator Trees from Model Embeddings
Anja Reusch and Wolfgang Lehner

End-to-End Evaluation of a Spoken Dialogue System for Learning Basic Mathematics
Eda Okur, Saurav Sahay, Roddy Fuentes Alba and Lama Nachman

15:30 - 16:00 *Coffee Break 2*

16:00 - 16:15 *Closing Remarks*

Tracing and Manipulating Intermediate Values in Neural Math Problem Solvers

Yuta Matsumoto [1] **Benjamin Heinzerling** [2] **Masashi Yoshikawa** [1] **Kentaro Inui** [1,2]
[1]Tohoku University / Japan [2]RIKEN / Japan
yuta.matsumoto.q8@dc.tohoku.ac.jp, benjamin.heinzerling@riken.jp,
yoshikawa@tohoku.ac.jp, inui@ecei.tohoku.ac.jp,

Abstract

How language models process complex input that requires multiple steps of inference is not well understood. Previous research has shown that information about intermediate values of these inputs can be extracted from the activations of the models, but it is unclear where that information is encoded and whether that information is indeed used during inference. We introduce a method for analyzing how a Transformer model processes these inputs by focusing on simple arithmetic problems and their intermediate values. To trace where information about intermediate values is encoded, we measure the correlation between intermediate values and the activations of the model using principal component analysis (PCA). Then, we perform a causal intervention by manipulating model weights. This intervention shows that the weights identified via tracing are not merely correlated with intermediate values, but causally related to model predictions. Our findings show that the model has a locality to certain intermediate values, and this is useful for enhancing the interpretability of the models.

1 Introduction

Recent language models (LMs) can solve complex input such as math word problems (Saxton et al., 2019; Geva et al., 2020). To obtain the correct output from such complex (latent structured) inputs, it is necessary for multiple steps of inference via intermediate values. However, how LMs process their inputs and capture latent structure is still not well understood. In previous studies, Linzen et al. (2016) and Tran et al. (2018) showed that the neural models can capture some implicit hierarchical structure, but it is unclear where that information is encoded. Shibata et al. (2020) observed that in LMs trained with Dyck language and showed some activations are highly correlated with the depth of their syntactic tree. However, even if such features can be extracted, there is no guarantee that it is used by the model (Elazar et al., 2021; Lovering et al., 2021). Given these considerations, to better understand LM predictions for latent structured inputs, it is necessary to: (a) To find where information about intermediate values of the latent structured inputs is encoded. (b) To evaluate the impact of the features when the model makes predictions.

In this work, we introduce a method for analyzing the relationship between internal representations in Transformer (Vaswani et al., 2017)-based models and intermediate values of latent structured inputs by using simple math problems. We choose them as a formal language because their intermediate values of the latent (tree) structure are clear and continuous, and it is easy to investigate their relationship to the internal representation of the model. The intermediate value of $(154 - 38) - (290 - 67)$ can be clearly defined as 154, 290, $154 - 38 = 116$, and so on. we take up a Transformer model trained to solve math equations. An overview of our experiments is shown in Fig. 1. First, we search which directions of internal representations are highly correlated with intermediate values in equations by PCA (**tracing**) to find where the information about intermediate values is encoded. We find some directions correlate very well with the intermediate values. Second, we observe how the model prediction changed when we manipulate the weights along its direction (**manipulation**) to conduct a causal intervention. The result of this experiment suggests that some directions of them are indeed used by the model.

These two results show that a Transformer model has a locality to certain intermediate values, and it could help enhance the interpretability of the models. Our contributions are as follows: (a)We show that intermediate values of equations are encoded in particular directions in internal representation. (b)We show that some features representing intermediate values are used during inference.

Proceedings of the 1st Workshop on Mathematical Natural Language Processing (MathNLP), pages 1 - 6
December 8, 2022 ©2022 Association for Computational Linguistics

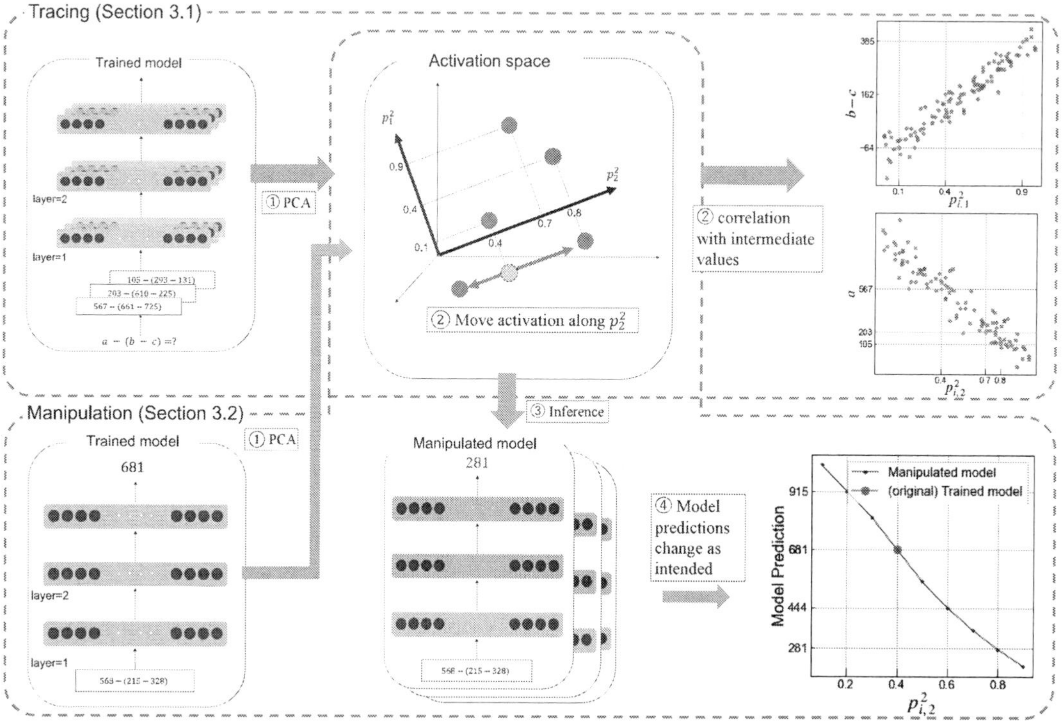

Figure 1: An overview of our methods. We find which directions obtained by PCA are correlated with intermediate values of the equations and how the model prediction changes when the weights of their directions are manipulated.

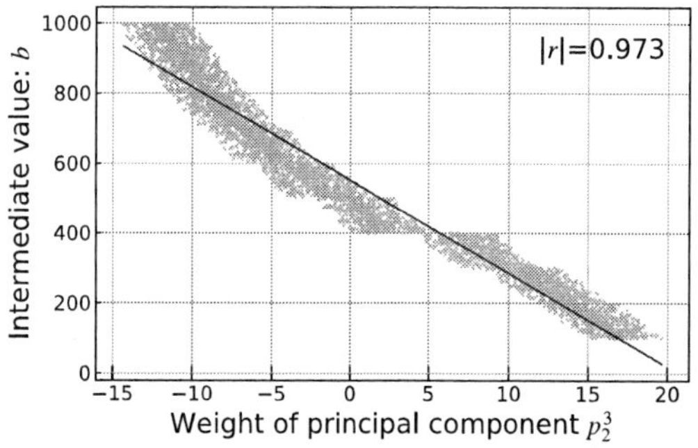

Figure 2: The relationship between $p_{i,2}^3$ and $R_i^j = b_i$ in the equation $a - (b - c)$. The correlation is very high.

2 Related Work

Intermediate values. Previous work has examined the representation of intermediate values in neural models. Linzen et al. (2016), Bowman et al. (2015) and Tran et al. (2018) found that LMs capture implicit hierarchical structures to some extent, e.g., when performing logical inference over formal languages. Closest to this work are Shibata et al. (2020), who trained LMs on the Dyck language and observed hidden units that are highly correlated with nesting depth. In contrast to their work, we analyze representations of more complex inputs, i.e., equations, and also manipulate these representations to understand the impact of correlated activations on model predictions.

Numeracy Geva et al. (2020) have shown that they can reach the state-of-the-art performance of numerical reasoning by using large pre-trained LM. Several studies have shown that a Transformer model can solve more complex problems such as linear algebra and elementary mathematics to some extent (Charton, 2021; Saxton et al., 2019). Based on their findings, we use simple mathematical equations as problems that can be solved by a Transformer model in this study.

3 Experiments

We conduct two types of experiments. First we trace the representation of intermediate values in the model. As a result we find directions in activation space that are highly correlation with intermediate values. Then we manipulate activations along these directions and observe if model predictions change as expected.

As neural math problem solving model, we train

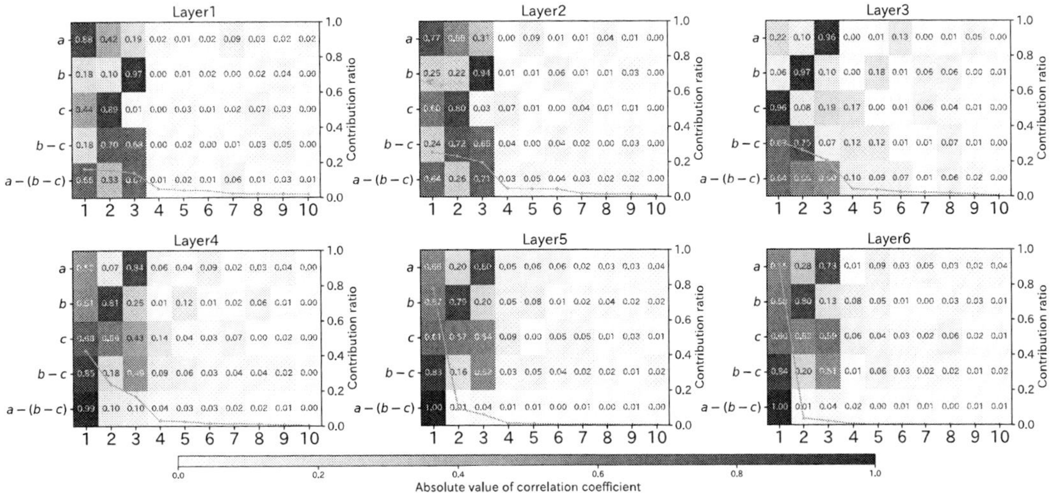

Figure 3: Correlations between each principal component and intermediate values, for all layers. Each cell represents the absolute value of the correlation coefficient between the weights of k-th principal component (column) and the intermediate values (row). The orange line shows the contribution ratio of each principal component.

a 6-layer Transformer using the settings by Sajjad et al. (2021) on synthetic data. We generate 200k equations involving up to five steps of addition or subtraction of integers between 1 and 1000, e.g., $(154 - 38) - (290 - 67)$. Following Geva et al. (2020) inputs are split into digits, e.g., "123" is tokenized into $1, \#\#2, \#\#3$. Model predictions are obtained via linear regression on the final layer's [CLS] token representation. After training on 190k equations we evaluate the model on 10k equations and obtain a regression score of $R^2 = 0.9988$, i.e., the model solves the equations almost perfectly.

3.1 Tracing intermediate values

Method. We describe our method for tracing the representation of intermediate values in model activations. First, we reduce the dimensionality of the activations at each model layer. Let h_j^l be the layer activations of the j-th word in a hidden layer l. Given an input of length n, we concatenate all token representations in layer l, obtaining the layer representation $H^l = h_1^l \oplus h_2^l \oplus \cdots \oplus h_n^l$ and fit a PCA to obtain the top 10 principal components $p_k^l, k \in [1, ..., 10]$. Applying this PCA to instance i yields the 10-dimensional representation $p_{i,k}^l$. Our hypothesis is that the intermediate values are encoded by one or more of the principal components. Intuitively, we assume that a principle component encodes an intermediate value if the magnitude of model activation in this direction correlates with the magnitude of the intermediate values. To test this hypothesis, we measure the correlation $\mathbf{corr}(R_i^j, p_{i,k}^l)$ between the value of the intermediate values R_i^j and the magnitude of principal component k in the representation $p_{i,k}^l$. Finally, we obtain **most-correlated direction** $\hat{p}_k^l(R^j) := \mathbf{argmax}_k(\mathbf{corr}(R_i^j, p_{i,k}^l))$. If this correlation is high, we conclude that the intermediate value is encoded in that direction.

Results. We trace intermediate values for the equation pattern $a - (b - c)$. For example, Fig. 2 shows a strong correlation of 0.973 between the intermediate value b and its most-correlated direction p_2^3. After measuring the correlation of each intermediate value and each of the top 10 principal components, we plot all correlations in Fig. 3. Overall, most-correlated directions show high correlations with intermediate values with moderate contribution ratio up to the 3rd layer, which we take as evidence that the model encodes intermediate values along these directions.

3.2 Manipulating intermediate values

Method. So far, we found correlations between intermediate values and directions in activation space. However, such correlations do not necessarily mean that these directions determine model predictions. To test if the directions we found actually influence model predictions, we perform causal interventions by *manipulating* activations. Concretely, we manipulate activations along principal

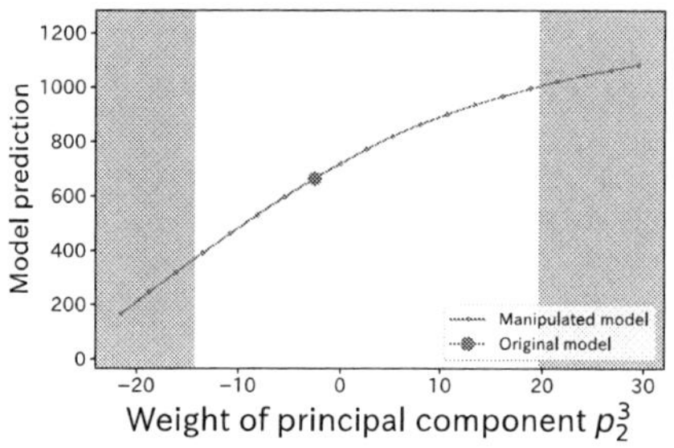

(a) Changes of model predictions as a function of weight of p_2^3.

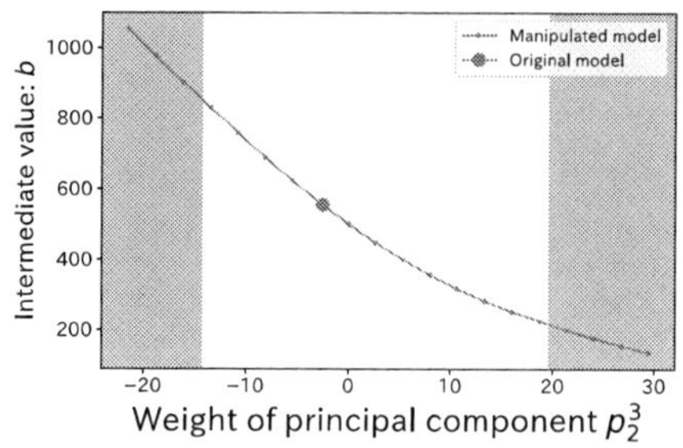

(b) The intermediate value b as a function of weight of p_2^3.

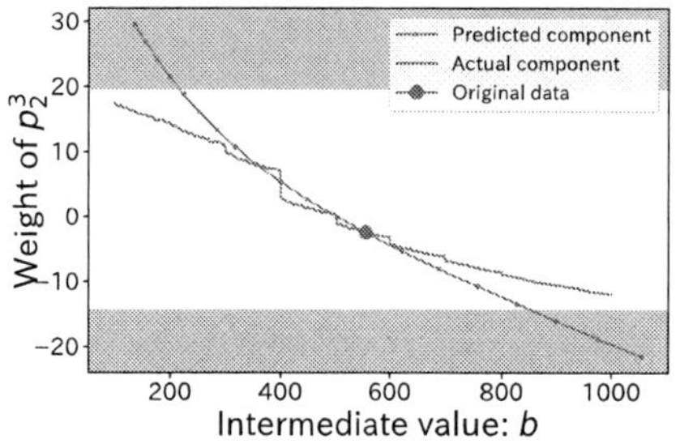

(c) Predicted and actual weights of most-correlated direction $\hat{p}_2^3(b)$ as a function of the intermediate value b.

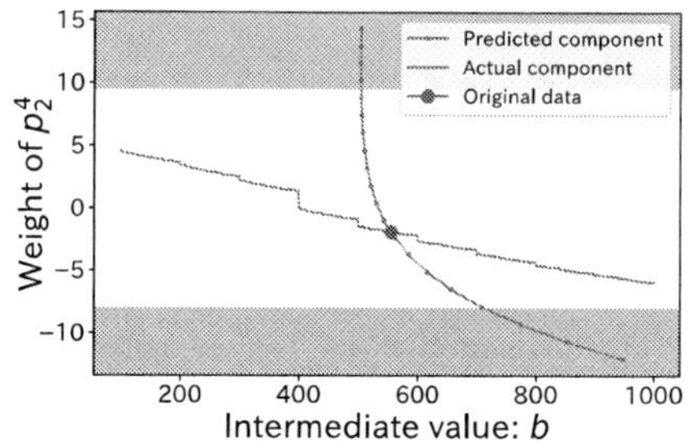

(d) Predicted and actual weights of most-correlated direction $\hat{p}_2^4(b)$ as a function of the intermediate value b.

Figure 4: The results of manipulation. The weights of the shaded areas do not appear in the dataset.

components and observe changes in model predictions, as shown in Fig. 1. Formally, we transform layer representation H^l (see §3.1) into $H^{l\prime}$, by increasing or decreasing its projection onto the principal component p_k^l by a factor of r:

$$H^{l\prime} \leftarrow H^l + (r-1)\left({p_k^l}^\top H^l\right) p_k^l \qquad (1)$$

Intuitively, increasing r moves H^l along p_k^l.

If a most-correlated direction $\hat{p}_k^l(R^j)$ indeed encodes the intermediate value R^j, it should be possible to manipulate activations in a way that corresponds to changing R^j. For example, if the model prediction given the input $43-(50-20)$ changes from the 13 to 19, this difference is consistent with changing the first input term from 43 to 49. By manipulation factors r of a particular most-correlated direction, observing model predictions, and calculating corresponding intermediate values, we obtain data for fitting a function from intermediate values to manipulation factors r. That is, we learn to manipulate activations in a way that corresponds to changing a particular intermediate value. To assess the fidelity of this manipulation, we change input terms and compare *actual* activation changes along the most-correlated direction $\hat{p}_k^l(R^j)$ to the factor r *predicted* by our fitted function.

Results. Using the input $617-(555-602)$ and the intermediate value $b=555$ as example, we find its most-correlated direction $\hat{p}_2^3(b)$, as described in §3.1. By manipulating activations along p_2^3, model predictions change from the original 664 to results ranging from ca. 200 to 1000, as shown in Fig. 4(a). Calculating intermediate values b that are consistent with these model predictions, we obtain Fig. 4(b). By axis inversion we obtain a function from b to *predicted* manipulation factors r for component $p_{i,2}^3$. We compare these *predicted* component weights to the *actual* component weights observed under changed inputs $\{(617-(i-602)) | (100 \leq i < 1000)\}$ (Fig. 4(c)). Predicted and the actual weights of the most-correlated direction agree well (corr. 0.986, R^2 score 0.687), which we take as evidence that $\hat{p}_2^3(b)$ encodes the intermediate value b and determines model predictions accordingly. Conversely, manipulation identifies most-correlated directions that are correlated but less used in prediction. The most-correlated direction $\hat{p}_2^4(b)$ has a high correlation of 0.81 with b, but predicted component weights show much less agreement with actual weights (corr. 0.802, R^2 score -1.06×10^4, Fig. 4(d)).

In conclusion, this case study showed how manipulations in activation space can find a causal connection to intermediate values.

4 Acknowledgments

This work was supported by JST CREST Grant Number JPMJCR20D2, Japan.

References

Samuel R. Bowman, Christopher D. Manning, and Christopher Potts. 2015. Tree-structured composition in neural networks without tree-structured architectures. In *Proceedings of the 2015th International Conference on Cognitive Computation: Integrating Neural and Symbolic Approaches - Volume 1583*, COCO'15, page 37–42, Aachen, DEU. CEUR-WS.org.

François Charton. 2021. Linear algebra with transformers.

Yanai Elazar, Shauli Ravfogel, Alon Jacovi, and Yoav Goldberg. 2021. Amnesic probing: Behavioral explanation with amnesic counterfactuals. *Transactions of the Association for Computational Linguistics*, 9:160–175.

Mor Geva, Ankit Gupta, and Jonathan Berant. 2020. Injecting numerical reasoning skills into language models. In *Proceedings of the 58th Annual Meeting of the Association for Computational Linguistics*, pages 946–958, Online. Association for Computational Linguistics.

Tal Linzen, Emmanuel Dupoux, and Yoav Goldberg. 2016. Assessing the ability of LSTMs to learn syntax-sensitive dependencies. *Transactions of the Association for Computational Linguistics*, 4:521–535.

Charles Lovering, Rohan Jha, Tal Linzen, and Ellie Pavlick. 2021. Predicting inductive biases of pre-trained models. In *International Conference on Learning Representations*.

Hassan Sajjad, Fahim Dalvi, Nadir Durrani, and Preslav Nakov. 2021. On the effect of dropping layers of pre-trained transformer models.

David Saxton, Edward Grefenstette, Felix Hill, and Pushmeet Kohli. 2019. Analysing mathematical reasoning abilities of neural models. In *International Conference on Learning Representations*.

Chihiro Shibata, Kei Uchiumi, and Daichi Mochihashi. 2020. How LSTM encodes syntax: Exploring context vectors and semi-quantization on natural text. In *Proceedings of the 28th International Conference on Computational Linguistics*, pages 4033–4043, Barcelona, Spain (Online). International Committee on Computational Linguistics.

Ke Tran, Arianna Bisazza, and Christof Monz. 2018. The importance of being recurrent for modeling hierarchical structure. In *Proceedings of the 2018 Conference on Empirical Methods in Natural Language Processing*, pages 4731–4736, Brussels, Belgium. Association for Computational Linguistics.

Ashish Vaswani, Noam Shazeer, Niki Parmar, Jakob Uszkoreit, Llion Jones, Aidan N Gomez, Ł ukasz Kaiser, and Illia Polosukhin. 2017. Attention is all you need. In *Advances in Neural Information Processing Systems*, volume 30. Curran Associates, Inc.

Investigating Math Word Problems using Pretrained Multilingual Language Models

Minghuan Tan and **Lei Wang** and **Lingxiao Jiang** and **Jing Jiang**
School of Computing and Information Systems
Singapore Management University
{mhtan.2017,lei.wang.2019}@phdcs.smu.edu.sg,{lxjiang,jingjiang}@smu.edu.sg

Abstract

In this paper, we revisit math word problems (MWPs) from the *cross-lingual* and *multilingual* perspective. We construct our MWP solvers over pretrained multilingual language models using the sequence-to-sequence model with copy mechanism. We compare how the MWP solvers perform in cross-lingual and multilingual scenarios. To facilitate the comparison of cross-lingual performance, we first adapt the large-scale English dataset MathQA as a counterpart of the Chinese dataset Math23K. Then we extend several English datasets to bilingual datasets through machine translation plus human annotation. Our experiments show that the MWP solvers may not be transferred to a different language even if the target expressions share the same numerical constants and operator set. However, it can be better generalized if problem types exist on both source language and target language.

1 Introduction

How to use machine learning and NLP techniques to solve Math Word Problems (MWPs) has attracted much attention in recent years (Hosseini et al., 2014; Kushman et al., 2014; Roy et al., 2015; Ling et al., 2017; Wang et al., 2017a, 2018; Amini et al., 2019). Given a math problem expressed in human language, a MWP solver typically first converts the input sequence of words to an *expression tree* consisting of math operators and numerical values, and then invokes an executor (such as the *eval* function in Python) to execute the expression tree to obtain the final numerical answer. Figure 1 shows an example math word problem, the correct expression tree, and the final answer.

Despite the relatively simple syntax of these expression trees, building MWP solvers is not a trivial task, and researchers have proposed various methods to tackle the different challenges of this problem such as statistical methods (Kushman

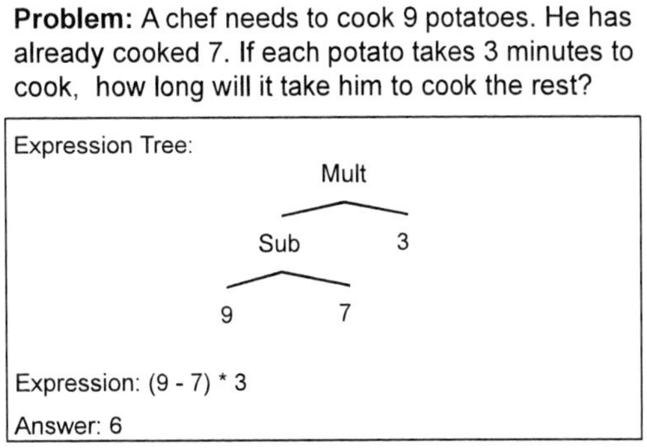

Figure 1: Example of an MWP and its expression tree.

et al., 2014; Roy et al., 2015), parsing-based methods (Shi et al., 2015) and generation-based methods (Wang et al., 2018; Xie and Sun, 2019). However, an aspect that has been largely overlooked is cross-lingual and multilingual MWP solving, i.e., whether a MWP solver trained on one human language can still work on another human language, or whether a MWP solver trained on multiple human languages together is more effective than a solver trained on only one language. We believe this is an interesting aspect to study for the following reasons. First, in cognitive science, people have long studied the relationship between humans' numerical processing abilities and language abilities, and found that on the one hand, the two are largely independent (Xu and Spelke, 2000), but on the other hand, "acquiring and mastering symbolic representations of exact quantities critically depends on language and instruction" (Van Rinsveld et al., 2015). It is therefore also intriguing to study whether machines separately acquire arithmetic and language abilities. Second, with pre-trained large-scale multilingual language models such as mBERT (Devlin et al., 2019) and XLM-R (Conneau et al., 2020), which presumably project different human languages into a common embedding space, we have seen some success in cross-lingual NLP tasks such as XNLI (Conneau et al., 2018) and

Proceedings of the 1st Workshop on Mathematical Natural Language Processing (MathNLP), pages 7 - 16
December 8, 2022 ©2022 Association for Computational Linguistics

MLQA (Lewis et al., 2020) in both zero-shot and few-shot settings (Wu and Dredze, 2019; Conneau et al., 2020). It is therefore reasonable to expect that for MWP solving, there is the possibility of transferring machine's capability of MWP solving from one language to another by leveraging these pre-trained multilingual language models.

In this paper, we conduct an empirical study to understand to what extent MWP solvers can work in cross-lingual and multilingual settings. Specifically, we ask the following questions: (1) Cross-lingual setting: Given a model trained with monolingual dataset, can the model solve MWPs over another language? (2) Multilingual setting: Can combining datasets of different languages further boost the performance for each language? (3) Can we identify some critical factors that may affect the results in (1) and (2)?

In order to empirically answer the questions above, we need multilingual MWP datasets, which are limited currently. We first use large scale datasets like Math23K (Wang et al., 2017b) and MathQA (Amini et al., 2019) as monolingual MWPs resource and further adapt MathQA to have the same operator set and expression style with Math23K. To better evaluate the models with parallel corpus, we extend some existing MWP datasets by translating them from English into Chinese. We then conduct three sets of experiments on the constructed datasets. We find that: (1) a cross-lingual MWP solver finetuned on one language cannot work on a second language, even if they are sharing the same decoding vocabulary, (2) a multilingual MWP solver may not boost performance for all the training languages but can improve those problems of similar types if one training language is close to the evaluation language, (3) combining (1) (2), we think for multilingual MWP solvers, despite language similarity, the performance relies heavily on domain similarity (problem types).

Our work makes the following contributions: (1) To the best of our knowledge, we are the first to study cross-lingual and multilingual MWP solving, and we empirically demonstrate that cross-lingual MWP solving is still difficult, but multilingual MWP solving is to some extent effective. (2) We discover that multilingual MWP solving is mostly effective for questions with similar problem types. (3) Our constructed datasets can help other researchers to further study cross-lingual and multilingual MWP solving.

2 Related work

Solving Math Word Problems (MWPs) has been attracting researchers since the emergence of artificial intelligence. STUDENT(Bobrow, 1964) is a rule-based math word problem solver which contains a pipeline that consists of heuristics for pattern transformation. Many researchers start with the fundamental problem types like addition and subtraction (Hosseini et al., 2014) or those that have only one single operator (Roy et al., 2015). Roy and Roth (2015) look at problems that require multi-steps using two or more operators. The question types of MWPs are also expanding. Rather than focusing on problems that need only one variable, Kushman et al. (2014) propose a dataset ALG514 which includes problems with a system of equations. With the development of deep learning, there has been a demand for large-scale datasets with more variations. Dolphin18K (Huang et al., 2016) is a large-scale dataset that is more than 9 times of the size of previous ones, and contains many more problem types. Math23K (Wang et al., 2017a) contains math word problems for elementary school students in Chinese language and is crawled from multiple online education websites. MathQA (Amini et al., 2019) is a new large-scale, diverse dataset of 37k multiple-choice math word problems in English and each problem is annotated with an executable formula using a new operation-based representation language. HMWP (Qin et al., 2020) contains three types of MWPs: arithmetic word problems, equations set problems, and non-linear equation problems.

Various approaches have been proposed to solve MWPs. Template-based approaches (Kushman et al., 2014; Zhou et al., 2015; Upadhyay et al., 2016; Huang et al., 2017) are widely adopted as numbers appeared in the expressions are usually sparse in the representation space and the expressions may fall into the same category. More recently, the community is also paying more attention to train a math solver by fine-tuning pretrained language models. For example, EPT (Kim et al., 2020) adopts ALBERT (Lan et al., 2020) as the encoder for its sequence-to-sequence module.

The monolingual performance gains achieved recently have not been evaluated from cross-lingual and cross-domain perspectives. Therefore, we decide to revisit MWPs using current SOTA pre-trained multilingual language models to construct a competitive math solver and conduct experiments over various bilingual evaluations.

3 Preliminaries

3.1 The MWP Solver Task

We first formally define the task of building MWP solvers. Given a math word problem with n words $W = (w_1, w_2, \ldots, w_n)$, and k numerical values $N = (n_0, n_1, \ldots, n_k)$, the model needs to generate a flattened tree representation using operators from permitted operator set $\mathcal{O}$ and numerical values from constants $\mathcal{C}$ and N. The generated tree should be able to be evaluated via some compiler and executor to return a numerical value.

3.2 Solution Framework

A MWP solver needs to generate executable code for a target programming language to be evaluated by an executor compiled for the programming language.

Our MWP solver is built upon a sequence-to-sequence model with copy mechanism (Gu et al., 2016). Specifically, we use a pretrained multilingual model as the encoder to get contextualized representations of math word problems. Due to the word piece tokenizer, the encoded context is not well-aligned to original input words. We choose to map these word pieces back to input words through mean pooling. Then we pass the mean pooled word representations to a bidirectional LSTM. Finally, we use a LSTM decoder with copy mechanism, which takes in the last decoded word vector and intermediate reading states, to predict the next token one by one. When the decoding finishes, we are expecting to get a linear tree representation. We attach the full model details in Section A.

Given the decoded tree representation, we first convert the generated linear tree representation into a piece of python expression with basic operators (+,-,*,/,**), then use the built-in function *eval* in Python to execute the generated code.

3.3 Existing Datasets

We use two large-scale datasets for this cross-lingual research. One is Math23K (Wang et al., 2017a) in Chinese and the other is MathQA (Amini et al., 2019) in English. Although the two datasets are similar in size and question types, there are still differences in terms with permitted operators and annotations.

Dataset	Problem Types	Size
AddSub (Hosseini et al., 2014)	Add	395
	Sub	
SingleOp (Roy et al., 2015)	Add	562
	Sub	
	Mult	
	Div	
MultiArith (Roy and Roth, 2015)	(Add, Sub)	600
	(Sub, Add)	
	(Add, Mult)	
	(Add, Div)	
	(Sub, Mult)	
	(Sub, Div)	

Table 1: Datasets which are focusing on specific problem types.

Math23K The dataset Math23K (Wang et al., 2017a) contains math word problems for elementary school students in *Chinese* (zh) and is crawled from multiple online education websites. The dataset focuses on arithmetic problems with a single-variable and contains 23,161 problems labeled with structured equations and answers.

MathQA The dataset is a new large-scale, diverse dataset of 37k multiple-choice math word problems in *English* (en). Each question is annotated with an executable formula using a new operation-based representation language (Amini et al., 2019). It covers multiple math domain categories. To make MathQA a comparable counterpart with Math23K, we choose to filter those solvable problems with shared permitted operators from MathQA to create an adapted MathQA dataset.

Other datasets focusing on specific problem types These datasets are smaller in size but more focused on specific problem types. We follow the dataset naming conventions from MAWPS (Koncel-Kedziorski et al., 2016).

Specifically, AddSub (Hosseini et al., 2014) covers arithmetic word problems on addition and subtraction for third, fourth, and fifth graders. Its problem types include combinations of additions, subtractions, one unknown equation, and U.S. money word problems. SingleOp (Roy et al., 2015) is a dataset with elementary math word problems of single operation. MultiArith (Roy and Roth, 2015) includes problems with multiple steps which we listed all the seven types in Table 1. These datasets are all in *English* (en). We will illustrate how we extend them into bilingual datasets in Section 4.2.

4 Cross-lingual and Multilingual MWP Solvers

In this work, as we are focusing on the cross-lingual and multilingual properties of MWPs, we need to train separate MWP solvers using different datasets. Our cross-lingual MWP solver will be trained using one language but evaluated using another. Our multilingual MWP solver can be trained on all languages available and evaluated separately. To suffice these goals, it would be better if the problems in different languages have comparable properties. Since we are using pretrained multilingual language models as the sequence embedder of the encoder, all the languages can be projected into a shared representation. However, the candidate datasets also need to share a common operator set and numerical constants to make the decoding process consistent. But some of the categories from MathQA do not exist on Math23K or one of the operators is not in our permitted set. Therefore, we need to adapt MathQA as a counterpart of Math23K sharing the same decoding vocabulary, including operators and constants.

4.1 Adaptation of MathQA

We adapt MathQA by doing the following:

1) We notice that the annotated formulas in MathQA are function calls of predefined functions which can be converted into a tree using an abstract syntax tree (AST) parser.
2) To be consistent with Math23K, which covers only basic arithmetic operators like addition (Add), subtraction (Sub), division (Div), multiplication (Mult) and exponentiation (Pow), we keep only functions in MathQA that can be expressed in such operators. For example, $volume\_sphere(r)$ from MathQA equalizes to $\frac{4}{3}\pi r^3$ and is adapted using the method shown in Figure 2. Formulas containing operators not used in Math23K, like *sine* and *permutation*, are not considered in this work. A full list of adapted operators can be found in Table 6 of Appendix A.
3) Upon constructing the trees using permitted operators, we evaluate each sample to verify its correctness against its ground-truth answer. Those cases that fail to get the correct answer are not considered in this work.

After the adaptation, we get the adapted MathQA dataset of solvable problems with comparable sizes and question types to Math23K. For Math23K, we further sample a development set of size 1000 from

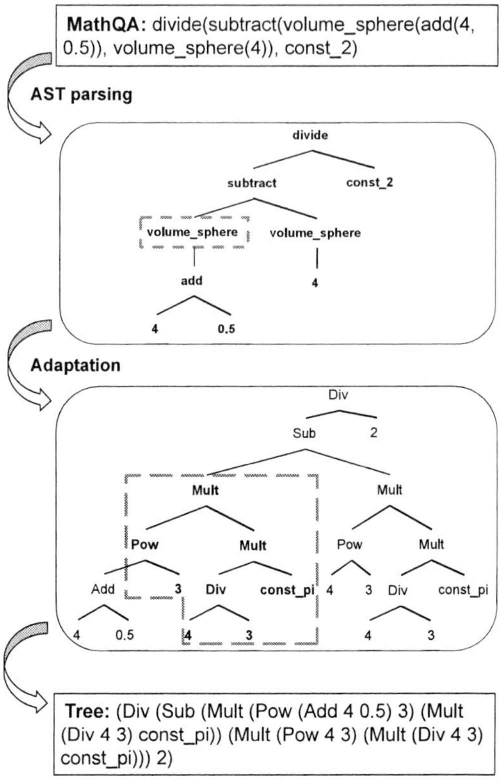

Figure 2: Adaptation of MathQA to Math23K. The part highlighted with dashed lines shows the adaptation of the function *volume_sphere*.

	Math23K		MathQA	
	w/o Pow	w/ Pow	w/o Pow	w/ Pow
Train	21,107	21,161	15,302	16,645
Dev	995	1,000	2,263	2,479
Test	999	1,000	1,532	1,653

Table 2: Statistics of different splits for Math23K and the adapted MathQA.

its training set. Considering the operator Pow has only several training and evaluating instances on Math23K, we separate them with others to make a fairer adaptation of MathQA to Math23K. We show the statistics of both Math23K and the adapted MathQA in Table 2. In this work, all the experiments will be conducted on the dataset marked with w/o Pow.

4.2 Zero-shot Cross-lingual Evaluation Datasets

To test cross-lingual transferability of MWP solvers, we make use of problem-type-specific datasets discussed in Section 3.3 as evaluation

Dataset	AddSub	SingleOp	MultiArith
Problem Types	addition, subtraction	single operation	multi-step
En	Keith has 20 books . Jason has 21 books . How many books do they have together ?	Lisa flew 256 miles at 32 miles per hour. How long did Lisa fly?	A chef needs to cook 9 potatoes. He has already cooked 7. If each potato takes 3 minutes to cook, how long will it take him to cook the rest?
Zh	基思有20本书。杰森有21本书。他们总共有多少本书?	丽莎以每小时32英里的速度飞行了256英里。丽莎飞了多长时间?	厨师需要煮9个土豆。他已经煮了7个了。如果每个土豆煮3分钟，剩下的他要煮多久?
Size	395	562	600

Table 3: Examples from each dataset used for zero-shot cross-lingual evaluation.

datasets, including AddSub (Hosseini et al., 2014), SingleOp (Roy et al., 2015) and MultiArith (Roy and Roth, 2015). To extend these datasets for cross-lingual evaluation, we use online machine translation APIs to translate them into Chinese and further manually refine the translations to be more native. For each dataset, we list an example in Table 3, in both English (En) and Chinese (Zh).

5 Template-based Contrastive Training

Math word problems can be categorized by expression templates if we replace numerical values of expressions with a special token. Such templates have been adopted for supervision in other math solver approaches like (Wang et al., 2018) and (Xie and Sun, 2019). Different from these methods, we don't use templates directly for supervision but make an assumption that problems sharing the same template are closer with each other from the point view of arithmetics, regardless of the surface forms of languages and descriptions.

To make use of this assumption, we introduce inter-language template-base contrastive training into our training process. Specifically, we first group math word problems based on their templates. During training, we pair each problem with a random sample from a different language sharing the same template.

As the representation learned by the encoder in Section A is $\mathbf{M}$, we use its maxpooling with normalization as the latent representation for each problem and its positive sample, denoted as $\mathbf{z}$ and $\mathbf{z}^+$ respectively. Then, we conduct a batch-level contrastive training similar to SimCLR (Chen et al., 2020) and use the NT-Xent loss (the normalized temperature-scaled cross entropy loss) as the following:

$$\mathcal{L} = -\log \frac{\exp(\langle \mathbf{z}, \mathbf{z}^+ \rangle / \tau)}{\exp(\langle \mathbf{z}, \mathbf{z}^+ \rangle / \tau) + \sum_{j=1}^{N-1} \exp(\langle \mathbf{z}, \mathbf{z}_j^- \rangle / \tau)}, \quad (1)$$

where $\langle \cdot, \cdot \rangle$ is the inner product of the two vectors and the batch size is N.

It's worth noting that the distribution of templates is highly skewed. In our experiments, we further consider two settings: (1) **CL**, contrastive learning, when a problem doesn't have a candidate with the same template from another language, it contrasts with itself. (2) **CL + TC**, contrastive learning with template constraint, we only use those problems which have at least one sample from another language.

Our contrastive learning approach differs with that of Li et al. (2022) in the following ways: (1) our method is focusing on cross-lingual setting that each pair of examples come from different languages, (2) we use batch-level contrastive training in consist with SimCLR.

There are also other works making use of latent representations of math word problems to enhance generalization ability of math solvers. For example, Liang and Zhang (2021) designed a teacher module to make the latent vector to match the correct solution rather than its variations.

6 Experiments

6.1 Experiment Setup

Evaluation metrics: The model is expected to be a math problem solver, so the generated expressions should be executable by a specific compiler and executor. During evaluation, each problem is counted as solved if the absolute error rate for the executed value and the target value is lower than a predefined threshold. In our experiments, we choose $1e^{-4}$ as the threshold. The final evaluation

Model	Test		Zero-shot					
	Math23K	MathQA	AddSub		SingleOp		MultiArith	
	zh	en	zh	en	zh	en	zh	en
mBERT-zh	**76.5**	3.3	30.9	10.4	66.0	32.7	**51.2**	15.7
mBERT-en	0.5	77.9	2.8	6.1	5.0	10.5	5.0	3.2
XLM-R-xl	75.5	**79.0**	**39.0**	21.3	67.4	40.4	44.7	**18.3**
mBERT-xl	76.3	**79.0**	35.2	**24.1**	**69.8**	**41.6**	45.0	16.0

Table 4: Comparisons of different cross-lingual models over Test set and zero-shot datasets.

metric is the accuracy of solved problem against all the problems.

Methods to be compared: We empirically compare the following cross-lingual: (1) **mBERT-zh** is using original multilingual BERT (Devlin et al., 2019) but trained over Math23K only; (2) **mBERT-en** is using original multilingual BERT (Devlin et al., 2019) but trained over the adapted MathQA only, and multilingual methods: (1) **mBERT-xl** is using original multilingual BERT (Devlin et al., 2019) but trained by mixing Math23K and the adapted MathQA; (2) **XLM-R-xl** is using XLM-R (Conneau et al., 2020) but trained by mixing Math23K and the adapted MathQA.

Other experiment settings: We choose to use multi-lingual BERT (mBERT) (Devlin et al., 2019) for cross-lingual training. We train our models using one Nvidia 2080ti and a batch size of 160. The learning rate is set to $3e^{-5}$ with a scheduler supporting polynomial decay. The training lasts for at most 150 epochs and will stop after 30 epochs if no improvement is observed.[1]

6.2 Results

We list experiment results of all the methods in Table 4.

Cross-lingual MWP Solver The first research question we want to answer is to what extent a MWP solver trained on one language can work on another language, with the help of pre-trained multilingual language models. Table 4 shows that the MWP solvers trained using either Math23K (**mBERT-zh**) or MathQA (**mBERT-en**) have achieved impressive performance when tested in the same language. However, the performance over a different language drops drastically and is almost negligible. In a word, the MWP solver is almost non-transferable when it is trained on one language but evaluated over a second with the same operator set.

Multilingual MWP Solver The second research question we want to answer is whether training a MWP solver on multiple languages helps improve its effectiveness compared with training on a single language. We can see that mixing two languages to train can give us a more language-agnostic model as the performance on Test split of both languages are competitive with monolingual cases. What's more, on the newly extended bilingual datasets, there are consistent improvements for most of the datasets, especially for the English language.

Considering the difficulty of problems, these bilingual evaluation datasets are closer to Math23K (primary school) than to MathQA (GRE or GMAT). Adding that mBERT-zh is also doing better than mBERT-en on English language, we suspect domain similarity is more important than language for MWP solvers.

Template-based Contrastive Training The last section of Table 5 shows how contrastive learning affects performance. Firstly, adding contrastive learning can further boost performance on the test set of both languages. There's a significant increase (3 points) for Math23K. However, in zero-shot evaluation settings, performance over English drops consistently. We suspect this might be caused by the diversity of templates on MathQA is much larger than that of Math23K.

Therefore, we further conduct a template-constrained experiment that ensures each template can be found on both languages. Due to the number of training cases are reduced, performance of the test sets also drop by a large margin. However, English problems over zero-shot setting benefit most from this experiment, which further verifies that

[1] `https://github.com/VisualJoyce/AnDuShu`

Model	Test		Zero-shot					
	Math23K	MathQA	AddSub		SingleOp		MultiArith	
	zh	en	zh	en	zh	en	zh	en
mBERT-xl	76.3	79.0	35.2	**24.1**	**69.8**	41.6	45.0	16.0
mBERT-xl + CL	**79.4**	**79.4**	39.5	19.2	62.6	29.9	**46.2**	10.8
mBERT-xl + CL + TC	72.4	49.5	**44.8**	19.2	67.4	**44.5**	45.5	**17.3**

Table 5: Performance of template-based contrastive training models over Test set and zero-shot datasets.

math word problems depend closely on the problem types of the training set.

7 Conclusion

In this paper, we revisit the math word problems using a generation-based method constructed over pretrained multilingual models. To assist analysis of cross-lingual properties of math solvers, we adopt two large-scale monolingual datasets and further adapts MathQA into the same annotation framework with Math23K. We also reuse earlier smaller datasets and upgrade them into bilingual datasets by machine translation and manual checking. Our experiments show that the MWP solvers may not be transferred to a different language even if the target expressions have the same operator set and constants. But for both cross-lingual and multilingual cases, it can be better generalized if problem types exist on both source language and target language. Problems considered to be easy by humans may still be hard for a math solver trained over the same language but from a different domain. This tells us that for math word problem solvers, it might be beneficial to consider balancing different question types and permitted operators during training.

References

Aida Amini, Saadia Gabriel, Shanchuan Lin, Rik Koncel-Kedziorski, Yejin Choi, and Hannaneh Hajishirzi. 2019. MathQA: Towards interpretable math word problem solving with operation-based formalisms. In *Proceedings of the 2019 Conference of the North American Chapter of the Association for Computational Linguistics: Human Language Technologies, Volume 1 (Long and Short Papers)*, pages 2357–2367, Minneapolis, Minnesota. Association for Computational Linguistics.

Daniel G. Bobrow. 1964. Natural language input for a computer problem solving system. Technical report, USA.

Ting Chen, Simon Kornblith, Mohammad Norouzi, and Geoffrey Hinton. 2020. A simple framework for contrastive learning of visual representations. In *Proceedings of the 37th International Conference on Machine Learning*, volume 119 of *Proceedings of Machine Learning Research*, pages 1597–1607. PMLR.

Alexis Conneau, Kartikay Khandelwal, Naman Goyal, Vishrav Chaudhary, Guillaume Wenzek, Francisco Guzmán, Edouard Grave, Myle Ott, Luke Zettlemoyer, and Veselin Stoyanov. 2020. Unsupervised cross-lingual representation learning at scale. In *Proceedings of the 58th Annual Meeting of the Association for Computational Linguistics*, pages 8440–8451, Online. Association for Computational Linguistics.

Alexis Conneau, Ruty Rinott, Guillaume Lample, Adina Williams, Samuel Bowman, Holger Schwenk, and Veselin Stoyanov. 2018. XNLI: Evaluating cross-lingual sentence representations. In *Proceedings of the 2018 Conference on Empirical Methods in Natural Language Processing*, pages 2475–2485, Brussels, Belgium. Association for Computational Linguistics.

Jacob Devlin, Ming-Wei Chang, Kenton Lee, and Kristina Toutanova. 2019. BERT: Pre-training of deep bidirectional transformers for language understanding. In *Proceedings of the 2019 Conference of the North American Chapter of the Association for Computational Linguistics: Human Language Technologies, Volume 1 (Long and Short Papers)*, pages 4171–4186, Minneapolis, Minnesota. Association for Computational Linguistics.

Jiatao Gu, Zhengdong Lu, Hang Li, and Victor O.K. Li. 2016. Incorporating copying mechanism in sequence-to-sequence learning. In *Proceedings of the 54th Annual Meeting of the Association for Computational Linguistics (Volume 1: Long Papers)*, pages 1631–1640, Berlin, Germany. Association for Computational Linguistics.

Mohammad Javad Hosseini, Hannaneh Hajishirzi, Oren Etzioni, and Nate Kushman. 2014. Learning to solve arithmetic word problems with verb categorization. In *Proceedings of the 2014 Conference on Empirical Methods in Natural Language Processing (EMNLP)*, pages 523–533, Doha, Qatar. Association for Computational Linguistics.

Danqing Huang, Shuming Shi, Chin-Yew Lin, and Jian Yin. 2017. Learning fine-grained expressions to solve math word problems. In *Proceedings of the 2017 Conference on Empirical Methods in Natural Language Processing*, pages 805–814, Copenhagen, Denmark. Association for Computational Linguistics.

Danqing Huang, Shuming Shi, Chin-Yew Lin, Jian Yin, and Wei-Ying Ma. 2016. How well do computers solve math word problems? large-scale dataset construction and evaluation. In *Proceedings of the 54th Annual Meeting of the Association for Computational Linguistics (Volume 1: Long Papers)*, pages 887–896, Berlin, Germany. Association for Computational Linguistics.

Bugeun Kim, Kyung Seo Ki, Donggeon Lee, and Gahgene Gweon. 2020. Point to the Expression: Solving Algebraic Word Problems using the Expression-Pointer Transformer Model. In *Proceedings of the 2020 Conference on Empirical Methods in Natural Language Processing (EMNLP)*, pages 3768–3779, Online. Association for Computational Linguistics.

Rik Koncel-Kedziorski, Subhro Roy, Aida Amini, Nate Kushman, and Hannaneh Hajishirzi. 2016. MAWPS: A math word problem repository. In *Proceedings of the 2016 Conference of the North American Chapter of the Association for Computational Linguistics: Human Language Technologies*, pages 1152–1157, San Diego, California. Association for Computational Linguistics.

Nate Kushman, Yoav Artzi, Luke Zettlemoyer, and Regina Barzilay. 2014. Learning to automatically solve algebra word problems. In *Proceedings of the 52nd Annual Meeting of the Association for Computational Linguistics (Volume 1: Long Papers)*, pages 271–281, Baltimore, Maryland. Association for Computational Linguistics.

Zhenzhong Lan, Mingda Chen, Sebastian Goodman, Kevin Gimpel, Piyush Sharma, and Radu Soricut. 2020. ALBERT: A lite BERT for self-supervised learning of language representations. In *International Conference on Learning Representations*.

Patrick Lewis, Barlas Oguz, Ruty Rinott, Sebastian Riedel, and Holger Schwenk. 2020. MLQA: Evaluating cross-lingual extractive question answering. In *Proceedings of the 58th Annual Meeting of the Association for Computational Linguistics*, pages 7315–7330, Online. Association for Computational Linguistics.

Zhongli Li, Wenxuan Zhang, Chao Yan, Qingyu Zhou, Chao Li, Hongzhi Liu, and Yunbo Cao. 2022. Seeking patterns, not just memorizing procedures: Contrastive learning for solving math word problems. In *Findings of the Association for Computational Linguistics: ACL 2022*, pages 2486–2496, Dublin, Ireland. Association for Computational Linguistics.

Zhenwen Liang and Xiangliang Zhang. 2021. Solving math word problems with teacher supervision. In *IJCAI*, pages 3522–3528.

Wang Ling, Dani Yogatama, Chris Dyer, and Phil Blunsom. 2017. Program induction by rationale generation: Learning to solve and explain algebraic word problems. In *Proceedings of the 55th Annual Meeting of the Association for Computational Linguistics (Volume 1: Long Papers)*, pages 158–167, Vancouver, Canada. Association for Computational Linguistics.

Jinghui Qin, Lihui Lin, Xiaodan Liang, Rumin Zhang, and Liang Lin. 2020. Semantically-aligned universal tree-structured solver for math word problems. In *Proceedings of the 2020 Conference on Empirical Methods in Natural Language Processing (EMNLP)*, pages 3780–3789, Online. Association for Computational Linguistics.

Subhro Roy and Dan Roth. 2015. Solving general arithmetic word problems. In *Proceedings of the 2015 Conference on Empirical Methods in Natural Language Processing*, pages 1743–1752, Lisbon, Portugal. Association for Computational Linguistics.

Subhro Roy, Tim Vieira, and Dan Roth. 2015. Reasoning about quantities in natural language. *Transactions of the Association for Computational Linguistics*, 3:1–13.

Shuming Shi, Yuehui Wang, Chin-Yew Lin, Xiaojiang Liu, and Yong Rui. 2015. Automatically solving number word problems by semantic parsing and reasoning. In *Proceedings of the 2015 Conference on Empirical Methods in Natural Language Processing*, pages 1132–1142, Lisbon, Portugal. Association for Computational Linguistics.

Shyam Upadhyay, Ming-Wei Chang, Kai-Wei Chang, and Wen-tau Yih. 2016. Learning from explicit and implicit supervision jointly for algebra word problems. In *Proceedings of the 2016 Conference on Empirical Methods in Natural Language Processing*, pages 297–306, Austin, Texas. Association for Computational Linguistics.

Amandine Van Rinsveld, Martin Brunner, Karin Landerl, Christine Schiltz, and Sonja Ugen. 2015. The relation between language and arithmetic in bilinguals: insights from different stages of language acquisition. *Frontiers in psychology*, 6:265.

Lei Wang, Yan Wang, Deng Cai, Dongxiang Zhang, and Xiaojiang Liu. 2018. Translating a math word problem to a expression tree. In *Proceedings of the 2018 Conference on Empirical Methods in Natural Language Processing*, pages 1064–1069, Brussels, Belgium. Association for Computational Linguistics.

Mingxuan Wang, Zhengdong Lu, Jie Zhou, and Qun Liu. 2017a. Deep neural machine translation with linear associative unit. In *Proceedings of the 55th Annual Meeting of the Association for Computational Linguistics (Volume 1: Long Papers)*, pages 136–145, Vancouver, Canada. Association for Computational Linguistics.

Yan Wang, Xiaojiang Liu, and Shuming Shi. 2017b. Deep neural solver for math word problems. In *Proceedings of the 2017 Conference on Empirical Methods in Natural Language Processing*, pages 845–854.

Shijie Wu and Mark Dredze. 2019. Beto, bentz, becas: The surprising cross-lingual effectiveness of BERT. In *Proceedings of the 2019 Conference on Empirical Methods in Natural Language Processing and the 9th International Joint Conference on Natural Language Processing (EMNLP-IJCNLP)*, pages 833–844, Hong Kong, China. Association for Computational Linguistics.

Zhipeng Xie and Shichao Sun. 2019. A goal-driven tree-structured neural model for math word problems. In *Proceedings of the Twenty-Eighth International Joint Conference on Artificial Intelligence, IJCAI-19*, pages 5299–5305. International Joint Conferences on Artificial Intelligence Organization.

Fei Xu and Elizabeth S Spelke. 2000. Large number discrimination in 6-month-old infants. *Cognition*, 74(1):B1–B11.

Lipu Zhou, Shuaixiang Dai, and Liwei Chen. 2015. Learn to solve algebra word problems using quadratic programming. In *Proceedings of the 2015 Conference on Empirical Methods in Natural Language Processing*, pages 817–822, Lisbon, Portugal. Association for Computational Linguistics.

A Method

In this section, we construct a generation-based MWP solver using a sequence-to-sequence model with copy mechanism. Our whole model can be visualized in modules through Figure 3. The detailed illustration for each module is given as following:

Encoder Our encoder is built upon a pretrained multilingual transformer, either BERT or XLM-R. Suppose our input word w_i is tokenized into word pieces $(x_{i1}, x_{i2}, \ldots)$ and let $\mathbf{h}_{ij} \in \mathbb{R}^{d_h}$ denotes the hidden vector produced by the pretrained model representing x_{ij}. We use average pooling to get the representation for the word w_i, denoted as $\mathbf{h}_i$. Then we feed this contextualized representation of the math word problem into a two-layer bidirectional LSTM. The output of this biLSTM is the encoder hidden states for decoding, denoted as $\mathbf{M} = (\mathbf{m}_0, \mathbf{m}_1, \ldots, \mathbf{m}_n)$.

Decoder We use a LSTM cell as the decoding cell to predict the next token. For each decoding step t, the cell will accept the embedding for previous word as input and output a decoder state $\mathbf{s}^t \in \mathbb{R}^{d_s}$. Most of the numerical values in MWPs do not exist in the target vocabulary. Therefore, we need copy mechanism (Gu et al., 2016) to facilitate generation of numerical values during decoding. The copy scores are calculated as follows,

$$u_i^t = \sigma(\mathbf{m}_i^\intercal \mathbf{W}_c)\mathbf{s}^t \quad (2)$$

where $\mathbf{W}_c \in \mathbb{R}^{d_h \times d_s}$. However, the embedding of a copied token will be identical to an out-of-vocabulary token. To better capture the information from last decoding step, we use the copy score to further derive a state of selecting from source tokens, which is called *Selective Read.*

$$\mathbf{q}^t = \mathrm{softmax}(\mathbf{u}^t) \quad (3)$$

$$\mathbf{b}^t = \sum_i q_i^t \mathbf{m}_i \quad (4)$$

We use a bilinear attention to attentively read information from $\mathbf{M}$, getting the context vector $\mathbf{c}^t$, which is called *Attentive Read.*

$$v_i^t = \sigma(\mathbf{m}_i^\intercal \mathbf{W}_a \mathbf{s}^t + b) \quad (5)$$

$$\mathbf{d}^t = \mathrm{softmax}(\mathbf{v}^t) \quad (6)$$

$$\mathbf{c}^t = \sum_i d_i^t \mathbf{m}_i \quad (7)$$

where $\mathbf{W}_a \in \mathbb{R}^{d_h \times d_s}$.

From the problem definition, the target vocabulary is $\mathcal{V} = \mathcal{O} \cup \mathcal{C}$. The generation score for the next token is given by:

$$\mathbf{p}^t = \mathbf{W}_d^\intercal \mathbf{s}^t + b \quad (8)$$

where $\mathbf{W}_d \in \mathbb{R}^{d_s \times |\mathcal{V}|}$.

The state updating process for the decoding cell takes in a fused information of last word embedding $\mathbf{e}^t \in \mathbb{R}^{d_e}$, selective read state $\mathbf{b}^t$ and attentive read state $\mathbf{c}^t$.

$$\mathbf{s}^{t+1} = f(\mathbf{W}_s[\mathbf{e}^t, \mathbf{b}^t, \mathbf{c}^t], \mathbf{s}^t) \quad (9)$$

where $\mathbf{W}_s \in \mathbb{R}^{d_s \times (d_e + d_h + d_h)}$.

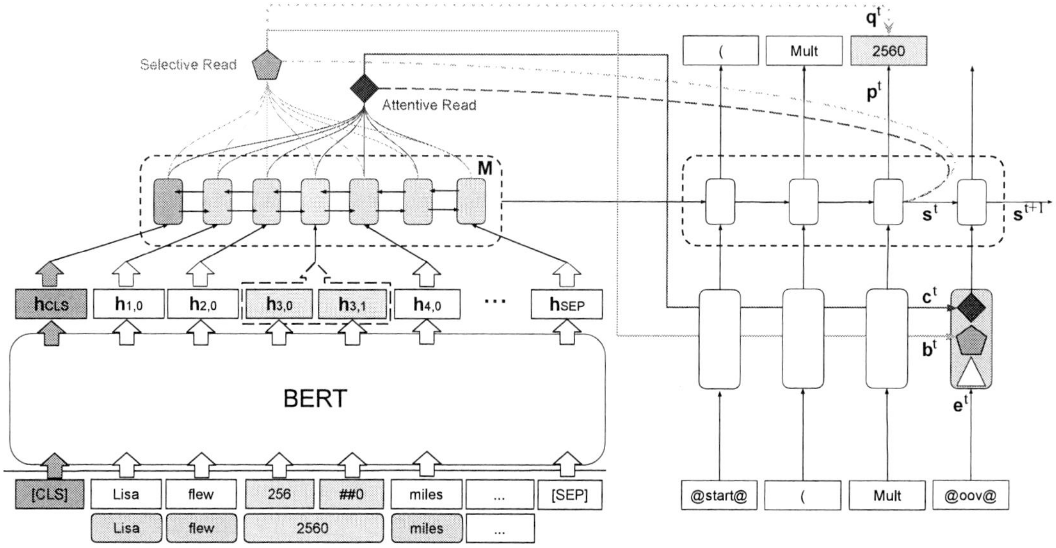

Figure 3: Sequence-to-sequence model with copy mechanism.

Adapted Operators	Filtered Operators
add, subtract, multiply, rectangle_area, divide, speed, power, negate, inverse, square_area, sqrt, square_edge_by_area, cube_edge_by_volume, volume_cube, surface_cube, square_perimeter, rectangle_perimeter, stream_speed, triangle_area, triangle_perimeter, surface_sphere, volume_sphere, rhombus_area, quadrilateral_area, volume_cylinder, circle_area, volume_cone, circumface, diagonal, volume_rectangular_prism, original_price_before_loss, original_price_before_gain, p_after_gain, square_edge_by_perimeter, negate_prob	floor, choose, min, tangent, sine, reminder, lcm, factorial, gcd, max, permutation, triangle_area_three_edges, surface_cylinder, rhombus_perimeter, surface_rectangular_prism, speed_in_still_water, log

Table 6: Operators that are adapted in MathQA.

Induced Natural Language Rationales and Interleaved Markup Tokens Enable Extrapolation in Large Language Models

Mirelle Bueno
University of Campinas

Carlos Gemmell
University of Glasgow

Jeffrey Dalton
University of Glasgow

Roberto Lotufo
University of Campinas
NeuralMind

Rodrigo Nogueira
University of Campinas
NeuralMind

Abstract

The ability to extrapolate, i.e., to make predictions on sequences that are longer than those presented as training examples, is a challenging problem for current deep learning models. Recent work shows that this limitation persists in state-of-the-art Transformer-based models. Most solutions to this problem use specific architectures or training methods that do not generalize to other tasks. We demonstrate that large language models can succeed in extrapolation without modifying their architecture or training procedure. Our experimental results show that generating step-by-step rationales and introducing marker tokens are both required for effective extrapolation. First, we induce a language model to produce step-by-step rationales before outputting the answer to effectively communicate the task to the model. However, as sequences become longer, we find that current models struggle to keep track of token positions. To address this issue, we interleave output tokens with markup tokens that act as explicit positional and counting symbols. Our findings show how these two complementary approaches enable remarkable sequence extrapolation and highlight a limitation of current architectures to effectively generalize without explicit surface form guidance. Code available at `https://github.com/MirelleB/induced-rationales-markup-tokens`

1 Introduction

The lack of compositional generalization of neural networks has been a long-standing limitation known for decades (Fodor and Pylyshyn, 1988; Schmidhuber, 1990; Marcus, 1998, 2018; Lake and Baroni, 2018; Liška et al., 2018; Keysers et al., 2019). This is often associated with their failure to extrapolate, i.e., the ability to work on sequences that are longer than those presented as training examples. Modern architectures such as the Transformer (Vaswani et al., 2017), which is the core component of state-of-the-art NLP models,

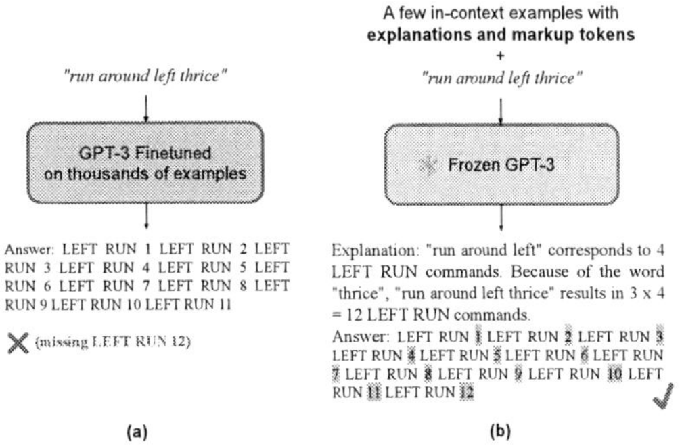

Figure 1: Answers produced by a GPT-3 model on the "length" split of the SCAN dataset when (a) fine-tuned on thousands of examples vs (b) induced via a few in-context examples to generate explanations and markup tokens (in yellow).

perform poorly on this class of problems (Bhattamishra et al., 2020; Nogueira et al., 2021; Wang et al., 2021; Pal and Baral, 2021; Welleck et al., 2021; Bogin et al., 2022; Finlayson et al., 2022; Mittal et al., 2021). In Figure 1-(a), we illustrate how recent large language models such as GPT-3 fail at this task, even when fine-tuned on thousands of examples.

Architectures and training methods that target this specific problem are often developed based on synthetic tasks whose creation rules are known (Das et al., 1992; Li et al., 2019b; Russin et al., 2019; Andreas, 2020; Liu et al., 2020a; Chen et al., 2020; Herzig and Berant, 2021; Shaw et al., 2021; Zhu et al., 2021). Thus, they resort to techniques such as augmenting the training data or biasing the model's architecture to internally represent these rules. However, improvements obtained on one compositional generalization benchmark do not transfer to others (Furrer et al., 2020), i.e., they lose their ability to be used as competitive general-purpose models in real tasks, as these can seldom be solved with a small set of rules.

We study the behavior of Transformer models

Proceedings of the 1st Workshop on Mathematical Natural Language Processing (MathNLP), pages 17 - 24
December 8, 2022 ©2022 Association for Computational Linguistics

and demonstrate that this problem is not due to an intrinsic limitation of their training algorithm. We show that inducing autoregressive models to rationalize before making a prediction (Wang et al., 2022; Zelikman et al., 2022) is not enough to extrapolate on long sequences: to solve it, we introduce markup tokens (Nogueira et al., 2021; Kim et al., 2021). The two general approaches together allow the models to achieve remarkable extrapolation generalization without requiring changes to the model or architecture. These findings provide evidence that general-purpose models have the ability to both improve their effectiveness and interpretability *at the same time*. The need to markup tokens also suggests there are fundamental issues that need to be addressed in the Transformer architecture, particularly the need for better positional representations. Thus, our study confirms and supports recent results from previous work that positional embeddings used in current state-of-the-art Transformer models cannot precisely track of token positions or perform precise counting (Liu et al., 2020b; Thawani et al., 2021; Press et al., 2022).

2 Related Work

A long list of architectures and training methods attempt to improve the extrapolation capabilities of deep learning models. For instance, some are specifically designed to solve only a handful of tasks (Singh, 1992; Kaiser and Sutskever, 2015; Kalchbrenner et al., 2015; Price et al., 2016; Andreas et al., 2016, 2017; Trask et al., 2018). Pre-trained word embeddings find it difficult to extrapolate to unseen numbers in training (Wallace et al., 2019). Alternatives to improving the extrapolation ability of neural models include building neural models with a pre-training corpus of numerical text (Geva et al., 2020) or using scientific notation to represent numbers (Zhang et al., 2020). Likewise, better numerical and compositional skills can be achieved by supplementing input texts with pre-computed numerical calculations (Andor et al., 2019) or explicitly assuming rules or mathematical equations from natural language texts (Liu et al., 2019; Li et al., 2019a; Zou and Lu, 2019a,b; Shi, 2020; Qiu et al., 2021). Many of these models are capable of adding numbers larger than those seen during training. In contrast, more general-purpose architectures fail to extrapolate on numerical tasks (Joulin and Mikolov, 2015; Dehghani et al., 2018; Schlag et al., 2019).

Our work derives from recent findings that show that inducing the model to generate explanations in natural language leads to better performance in a wide variety of tasks (Recchia, 2021; Fernandes et al., 2022; Wang et al., 2022; Zelikman et al., 2022; Nye et al., 2022; Katz et al., 2022; Zhou et al., 2022; Khot et al., 2022). In particular, the work proposed by (Zhou et al., 2022) achieves state-of-the-art results in the extrapolation of tasks involving symbolic manipulation, compositional generalization and numerical reasoning. Tasks are solved via few-shot learning applied to a large language model (e.g. text-davinci-002) in two main steps. The first step consists of reducing the question into sub-questions, then, in the second phase, a new interaction is made with the model, now solving sequentially the sub-questions generated in the previous step.

The results shown in Zhou et al. (Zhou et al., 2022) corroborate our intuition that explanations alone are not enough to achieve extrapolations. By inducing the model to generate explanations *and* markup tokens, *we provide evidence that compositional generalization can be achieved without sacrificing the general applicability on other tasks*, which is often a feature that is lost with architectural modifications.

However, a limitation of Zhou et al.'s and our method is that both require a programmatic post-processing step: Zhou et al. use a python script to convert the model output (e.g., `3*["LEFT"]`), which is in python notation, into the expected format of the final answer (e.g., `LEFT LEFT LEFT`); in our method, we programmatically remove the markup tokens from the final answer. We argue that the need to call an external script exposes a limitation in the current Transformer architecture, namely, that it cannot handle long sequences of repeated tokens.

3 Methodology

In this section, we describe our proposed method for inducing explanations and markup tokens using in-context learning with a few examples. We first create a prompt $ic||oc$ that concatenates in-context training examples ic with a test example oc. The ic examples consist of N triples of "Instruction", "Explanation" and "Output", i.e., $ic = \{(i_1^*, e_1^*, o_1^*), ..., (i_N^*, e_N^*, o_N^*)\}$. The test example oc is made of only the "Instruction" field. When we feed $ic||oc$ to a language model, it should

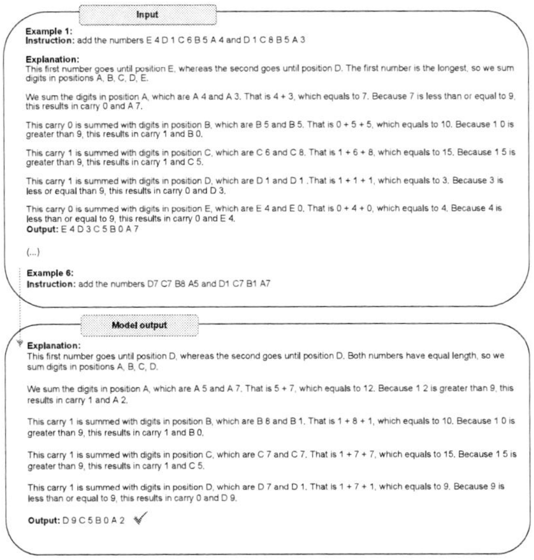

Figure 2: Example of a few-shot prompt and model completion for the addition task. First, a prompt composed of in-context (ic) samples are given, which are formed by $\{input, explanation, output\}$ triplets concatenated with an out-of-context (oc) test example that has only the "instruction" field. The model then completes the "explanation" and "output" fields from the test example as a result.

generate the remaining "Explanation" and "Output" fields for oc. Figure 2 illustrates the input prompt given to the model and the (correct) output given by the model.

We also interleave the tokens ic and oc with markup tokens that help the model to precisely identify the tokens in the input and output sequences (see Figure 1-(b) for an example). These tokens support the model in three ways: 1) They act as a form of working memory to indicate progress being made. 2) They act as sub-prompt anchors to inform the start of a known pattern. 3) They implicitly model a stopping condition should a certain amount of progress be reached. We programmatically include these markups in each test input and remove them from the output answers before comparing them with ground-truth ones.

Due to its few-shot nature, our method can be adjusted for different tasks. Likewise, our approach does not require any additional modifications to the language model such as pretraining or changes to the loss function.

4 Experimental Setup

We evaluate our method in two tasks that require extrapolation: 1) the length split of the SCAN data (Lake and Baroni, 2018) and 2) the addition of two numbers. In all experiments, we used the `text-davinci-002` model, available via a paid API provided by OpenAI. We report the accuracy of the test set.

4.1 SCAN

The SCAN synthetic dataset translates simple navigation commands into a sequence of actions (e.g., the input `jump thrice` results in the output `JUMP JUMP JUMP`). These commands are generated from the composition of a specific grammar, combining "primitive" commands such as `jump`, `walk`, `look`, `run` and `turn`; "modifiers" (`left`, `right`, `around`, `opposite`); repetition symbols like `twice`/`thrice`; "combiners" (`and`/`after`) that group two action sequences.

To construct the prompt, we generated nine in-context training examples, each made of three parts: an instruction, an explanation, and the desired output. The "Instruction" is a sequence of commands while the "Explanation" is a description, in natural language detailing the steps to generate the output. The "Output" corresponds to the expected answer to the instruction. In addition, in the output field, we inject markup tokens to delimit the end of a repeating sequence or sub-instruction. Therefore, to indicate each repetition of a given action, we use positive integers and at the end of a sequence of actions, we use the separator `||`. For example, for the input: `jump twice and walk twice`, we generate the output `JUMP 1 JUMP 2 || WALK 1 WALK 2`.

The target outputs of training examples have up to 22 actions. The test examples were drawn from the "Length" split provided by the authors.* This set has 3,920 examples whose target output varies between 24 and 48 actions. The instruction (input) of each test example is appended to the in-context training examples and the model is prompted to generate the "Explanation" and "Output" fields. Thus, since training examples are shorter than test ones, we are able to assess the compositional generalization of the model while extrapolating to larger unseen sequences. Due to the cost of using the GPT-3 API (approximately 0.10 USD per example),

*https://github.com/brendenlake/SCAN

Method	Acc.
Specialized Architectures	
Syntactic Attn. (Russin et al., 2019)	15.2
CGPS (Li et al., 2019b)	20.3
T5-base DUEL (Zhu et al., 2021)	45.0
LANE (Liu et al., 2020a)	100.0
NSSM (Chen et al., 2020)	100.0
SBSP (Herzig and Berant, 2021)	100.0
NQG (Shaw et al., 2021)	100.0
Synth (Nye et al., 2020)	100.0
General-purpose Architectures	
T5-base (Furrer et al., 2020)	14.4
T5-Large (Furrer et al., 2020)	5.2
T5-3B (Furrer et al., 2020)	3.3
T5-11B (Furrer et al., 2020)	2.0
GPT-3 Ada - fine-tuned	13.9
GPT-3 Curie - fine-tuned	6.4
GPT-3 Davinci - fine-tuned	8.2
Least-to-Most (Zhou et al., 2022)	99.7
—	
Ours (rationales only)	2.5
Ours (markups only)	22.5
Ours (rationales + markups, inverted prompt)	30.0
Ours (rationales + markups)	95.2

Table 1: Results on the "length" split of the SCAN dataset.

we evaluated the model on 400 randomly sampled examples from the test set.

4.2 Addition Task

Extrapolation abilities can also be tested with arithmetic tasks. For this, we built a prompt for the addition operation, where we present five in-context training examples with two numbers up to 5 digits and ask the model to generate the explanation and answer for a test set example made of numbers with 4 to 14 digits. We evaluate the model on 400 test samples automatically generated by the "balanced sampling" method from Nogueira et al. (Nogueira et al., 2021), which ensures that the set will have a roughly equal proportion of answers with d-digit numbers, with $d \in [4, 14]$.

We use a template similar to SCAN's to feed the in-context examples to the model. We manually generate the explanations for the training examples and inject markup tokens in the instructions and the target output. In the expected output, these tokens are used during the explanation steps. We illustrate in Figure 2 an example of a prompt followed by a completion of the model.

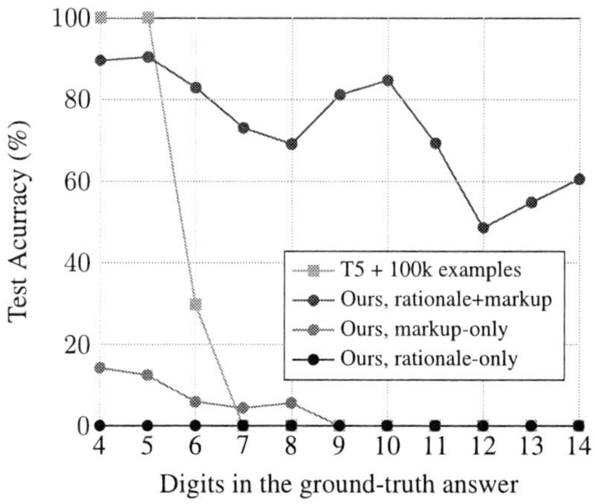

Figure 3: Test set accuracy in the addition task vs number of digits in the ground-truth answer.

5 Results

In Table 1 we show the results for the length split of the SCAN dataset. We see that specialized models like LANE, NSSM, and SBSP solve the compositional generalization proposed by SCAN, whereas generic architectures such as T5 (Raffel et al., 2020) or GPT-3 (Brown et al., 2020) fine-tuned on the task have poor performance.

We also show results for GPT-3's Ada (300M parameters), Curie (6B parameters) and Davinci (175B parameters) models fine-tuned on all 16,990 training examples of the SCAN dataset for 3 epochs. In these cases, we do not use in-context examples, explanations, or markup tokens. Our methodology of providing prompts with detailed explanations was shown to be more effective than finetuning on thousands of examples.

The same behavior is also observed in the addition task, as seen in Figure 3. Our approach with explanations and markup tokens (rationale + markup) shows that even with as few as 5 examples, the model can perform the task of adding numbers with more than 5 digits, reaching a performance of around 60% in numbers with up to 14 digits and an average accuracy of 73% considering all 400 examples in the test set.

We also investigated the performance of finetuning a general-purpose model on this task. We trained a T5-base with 100K samples on numbers with 2 to 5 digits per 10 epochs without adding explanations. We observe that the model reaches 100% accuracy with numbers of up to 5 digits, but fails to add numbers with more than 6 digits.

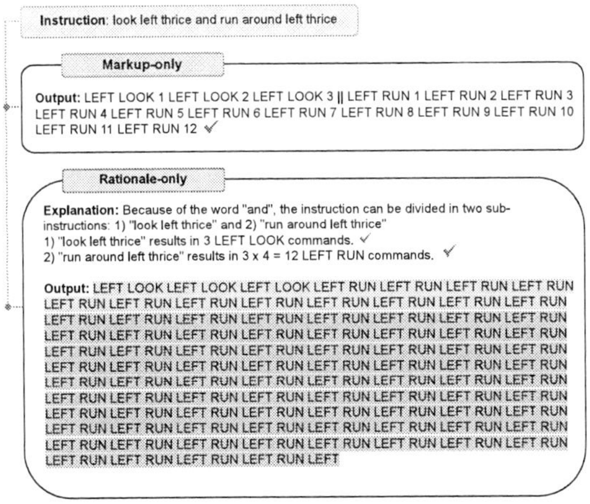

Figure 4: Model output differences between the markup-only and rationale-only approaches.

5.1 Ablation: Rationale-only vs. markup-only

We also investigate the impact of using explanation and markup tokens in isolation. We compare two scenarios: prompts without explanation (markups-only) and without markup tokens (rationales-only).

In Table 1, we see that the rationale-only and markup-only approaches have significantly lower test accuracy, demonstrating that it is not enough to explain how to solve the task, but it is also important to inject markup tokens. We believe that these tokens help the model generate repeated sequences of tokens.

In Figure 4, we provide qualitative evidence of this hypothesis: Without markup tokens, the model correctly generates the explanation but fails to finish the action sequence, therefore entering a loop.

5.2 Ablation: Inverted prompt

We also experimented with reversing the order in which the "explanations" and "outputs" fields are presented to the model. Therefore we provide the expected output first and then the explanation. The idea of this experiment was to verify if the order explanation followed by the output has an impact on the generation of the answer. In Table 1 we see that the performance drops from 95.2 to 30% (rationales + markups - inverted prompt). This empirical result agrees with the literature in terms that the model possibly processes the explanation before determining the final output.

6 Conclusion

In this work, we show how step-by-step rationales and positional markup tokens enable general-purpose architectures to extrapolate to sequences that are significantly longer than those provided as training examples. Rationales before the answer break down the problem into small executable chunks and markup tokens track the working progress as the output is generated. Importantly, we show how these methods are complementary and, when used together, enable remarkable extrapolation results on two synthetic tasks.

However, we note the use of markup tokens as a limitation of current models and subword tokenizers. Future models should be able to count tokens and keep track of individual tokens in long sequences without resorting to additional supporting tokens. As our qualitative analysis shows, most failure cases are due to one or two tokens generated incorrectly. We see the ability to automatically verify these errors, as proposed by Cobbe et al. (Cobbe et al., 2021), as a promising direction to improve the extrapolation capabilities of current models.

Acknowledgments

This research was partially funded by grants 2020/09753-5 and 2022/01640-2 from Fundação de Amparo à Pesquisa do Estado de São Paulo (FAPESP). This work is also supported by grant EP/V025708/1 from the Engineering and Physical Sciences Research Council. We also thank Google Cloud for credits to support this work. R Lotufo is partially supported by CNPq (The Brazilian National Council for Scientific and Technological Development) under grant 310828/2018-0.

References

Daniel Andor, Luheng He, Kenton Lee, and Emily Pitler. 2019. Giving BERT a calculator: Finding operations and arguments with reading comprehension. In *Proceedings of the 2019 Conference on Empirical Methods in Natural Language Processing and the 9th International Joint Conference on Natural Language Processing (EMNLP-IJCNLP)*, pages 5949–5954.

Jacob Andreas. 2020. Good-enough compositional data augmentation. In *Proceedings of the 58th Annual Meeting of the Association for Computational Linguistics*, pages 7556–7566, Online. Association for Computational Linguistics.

Jacob Andreas, Dan Klein, and Sergey Levine. 2017. Modular multitask reinforcement learning with policy sketches. In *International Conference on Machine Learning*, pages 166–175. PMLR.

Jacob Andreas, Marcus Rohrbach, Trevor Darrell, and Dan Klein. 2016. Learning to compose neural networks for question answering. In *Proceedings of the 2016 Conference of the North American Chapter of the Association for Computational Linguistics: Human Language Technologies*, pages 1545–1554.

Satwik Bhattamishra, Kabir Ahuja, and Navin Goyal. 2020. On the ability and limitations of transformers to recognize formal languages. In *Proceedings of the 2020 Conference on Empirical Methods in Natural Language Processing (EMNLP)*, pages 7096–7116.

Ben Bogin, Shivanshu Gupta, and Jonathan Berant. 2022. Unobserved local structures make compositional generalization hard. *arXiv preprint arXiv:2201.05899*.

Tom Brown, Benjamin Mann, Nick Ryder, Melanie Subbiah, Jared D Kaplan, Prafulla Dhariwal, Arvind Neelakantan, Pranav Shyam, Girish Sastry, Amanda Askell, et al. 2020. Language models are few-shot learners. *Advances in neural information processing systems*, 33:1877–1901.

Xinyun Chen, Chen Liang, Adams Wei Yu, Dawn Song, and Denny Zhou. 2020. Compositional generalization via neural-symbolic stack machines. *Advances in Neural Information Processing Systems*, 33:1690–1701.

Karl Cobbe, Vineet Kosaraju, Mohammad Bavarian, Jacob Hilton, Reiichiro Nakano, Christopher Hesse, and John Schulman. 2021. Training verifiers to solve math word problems. *arXiv preprint arXiv:2110.14168*.

Sreerupa Das, C. Giles, and Gordon Sun. 1992. Learning context-free grammars: Capabilities and limitations of a recurrent neural network with an external stack memory.

Mostafa Dehghani, Stephan Gouws, Oriol Vinyals, Jakob Uszkoreit, and Lukasz Kaiser. 2018. Universal transformers. In *International Conference on Learning Representations*.

Patrick Fernandes, Marcos Treviso, Danish Pruthi, André FT Martins, and Graham Neubig. 2022. Learning to scaffold: Optimizing model explanations for teaching. *arXiv preprint arXiv:2204.10810*.

Matthew Finlayson, Kyle Richardson, Ashish Sabharwal, and Peter Clark. 2022. What makes instruction learning hard? an investigation and a new challenge in a synthetic environment. *arXiv preprint arXiv:2204.09148*.

Jerry A Fodor and Zenon W Pylyshyn. 1988. Connectionism and cognitive architecture: A critical analysis. *Cognition*, 28(1-2):3–71.

Daniel Furrer, Marc van Zee, Nathan Scales, and Nathanael Schärli. 2020. Compositional generalization in semantic parsing: Pre-training vs. specialized architectures. *arXiv preprint arXiv:2007.08970*.

Mor Geva, Ankit Gupta, and Jonathan Berant. 2020. Injecting numerical reasoning skills into language models. In *Proceedings of the 58th Annual Meeting of the Association for Computational Linguistics*, pages 946–958.

Jonathan Herzig and Jonathan Berant. 2021. Span-based semantic parsing for compositional generalization. In *Proceedings of the 59th Annual Meeting of the Association for Computational Linguistics and the 11th International Joint Conference on Natural Language Processing (Volume 1: Long Papers)*, pages 908–921, Online. Association for Computational Linguistics.

Armand Joulin and Tomas Mikolov. 2015. Inferring algorithmic patterns with stack-augmented recurrent nets. *Advances in Neural Information Processing Systems*, 28:190–198.

Łukasz Kaiser and Ilya Sutskever. 2015. Neural GPUs learn algorithms. *arXiv preprint arXiv:1511.08228*.

Nal Kalchbrenner, Ivo Danihelka, and Alex Graves. 2015. Grid long short-term memory. *arXiv preprint arXiv:1507.01526*.

Uri Katz, Mor Geva, and Jonathan Berant. 2022. Inferring implicit relations with language models. *arXiv preprint arXiv:2204.13778*.

Daniel Keysers, Nathanael Schärli, Nathan Scales, Hylke Buisman, Daniel Furrer, Sergii Kashubin, Nikola Momchev, Danila Sinopalnikov, Lukasz Stafiniak, Tibor Tihon, et al. 2019. Measuring compositional generalization: A comprehensive method on realistic data. In *International Conference on Learning Representations*.

Tushar Khot, Harsh Trivedi, Matthew Finlayson, Yao Fu, Kyle Richardson, Peter Clark, and Ashish Sabharwal. 2022. Decomposed prompting: A modular approach for solving complex tasks. *arXiv preprint arXiv:2210.02406*.

Jeonghwan Kim, Giwon Hong, Kyung-min Kim, Junmo Kang, and Sung-Hyon Myaeng. 2021. Have you seen that number? investigating extrapolation in question answering models. In *Proceedings of the 2021 Conference on Empirical Methods in Natural Language Processing*, pages 7031–7037.

Brenden Lake and Marco Baroni. 2018. Generalization without systematicity: On the compositional skills of sequence-to-sequence recurrent networks. In *International conference on machine learning*, pages 2873–2882. PMLR.

Jierui Li, Lei Wang, Jipeng Zhang, Yan Wang, Bing Tian Dai, and Dongxiang Zhang. 2019a. Modeling intra-relation in math word problems with different functional multi-head attentions. In *Proceedings of the*

57th Annual Meeting of the Association for Computational Linguistics*, pages 6162–6167.

Yuanpeng Li, Liang Zhao, Jianyu Wang, and Joel Hestness. 2019b. Compositional generalization for primitive substitutions. In *Proceedings of the 2019 Conference on Empirical Methods in Natural Language Processing and the 9th International Joint Conference on Natural Language Processing (EMNLP-IJCNLP)*, pages 4293–4302, Hong Kong, China. Association for Computational Linguistics.

Adam Liška, Germán Kruszewski, and Marco Baroni. 2018. Memorize or generalize? searching for a compositional rnn in a haystack. *arXiv preprint arXiv:1802.06467*.

Qian Liu, Shengnan An, Jian-Guang Lou, Bei Chen, Zeqi Lin, Yan Gao, Bin Zhou, Nanning Zheng, and Dongmei Zhang. 2020a. Compositional generalization by learning analytical expressions. *Advances in Neural Information Processing Systems*, 33:11416–11427.

Qianying Liu, Wenyv Guan, Sujian Li, and Daisuke Kawahara. 2019. Tree-structured decoding for solving math word problems. In *Proceedings of the 2019 Conference on Empirical Methods in Natural Language Processing and the 9th International Joint Conference on Natural Language Processing (EMNLP-IJCNLP)*, pages 2370–2379.

Xuanqing Liu, Hsiang-Fu Yu, Inderjit Dhillon, and Cho-Jui Hsieh. 2020b. Learning to encode position for transformer with continuous dynamical model. In *International Conference on Machine Learning*, pages 6327–6335. PMLR.

Gary Marcus. 2018. Deep learning: A critical appraisal. *arXiv preprint arXiv:1801.00631*.

Gary F Marcus. 1998. Rethinking eliminative connectionism. *Cognitive psychology*, 37(3):243–282.

Sarthak Mittal, Sharath Chandra Raparthy, Irina Rish, Yoshua Bengio, and Guillaume Lajoie. 2021. Compositional attention: Disentangling search and retrieval. *CoRR*, abs/2110.09419.

Rodrigo Nogueira, Zhiying Jiang, and Jimmy Lin. 2021. Investigating the limitations of transformers with simple arithmetic tasks. *arXiv preprint arXiv:2102.13019*.

Maxwell Nye, Anders Johan Andreassen, Guy Gur-Ari, Henryk Michalewski, Jacob Austin, David Bieber, David Dohan, Aitor Lewkowycz, Maarten Bosma, David Luan, Charles Sutton, and Augustus Odena. 2022. Show your work: Scratchpads for intermediate computation with language models. In *Deep Learning for Code Workshop*.

Maxwell Nye, Armando Solar-Lezama, Josh Tenenbaum, and Brenden M Lake. 2020. Learning compositional rules via neural program synthesis. *Advances in Neural Information Processing Systems*, 33:10832–10842.

Kuntal Kumar Pal and Chitta Baral. 2021. Investigating numeracy learning ability of a text-to-text transfer model. In *Findings of the Association for Computational Linguistics: EMNLP 2021*, pages 3095–3101.

Ofir Press, Noah Smith, and Mike Lewis. 2022. Train short, test long: Attention with linear biases enables input length extrapolation. In *International Conference on Learning Representations*.

Eric Price, Wojciech Zaremba, and Ilya Sutskever. 2016. Extensions and limitations of the neural GPU. *arXiv preprint arXiv:1611.00736*.

Linlu Qiu, Peter Shaw, Panupong Pasupat, Paweł Krzysztof Nowak, Tal Linzen, Fei Sha, and Kristina Toutanova. 2021. Improving compositional generalization with latent structure and data augmentation. *arXiv preprint arXiv:2112.07610*.

Colin Raffel, Noam Shazeer, Adam Roberts, Katherine Lee, Sharan Narang, Michael Matena, Yanqi Zhou, Wei Li, and Peter J Liu. 2020. Exploring the limits of transfer learning with a unified text-to-text transformer. *Journal of Machine Learning Research*, 21:1–67.

Gabriel Recchia. 2021. Teaching autoregressive language models complex tasks by demonstration. *arXiv preprint arXiv:2109.02102*.

Jake Russin, Jason Jo, Randall C O'Reilly, and Yoshua Bengio. 2019. Compositional generalization in a deep seq2seq model by separating syntax and semantics. *arXiv preprint arXiv:1904.09708*.

Imanol Schlag, Paul Smolensky, Roland Fernandez, Nebojsa Jojic, Jürgen Schmidhuber, and Jianfeng Gao. 2019. Enhancing the transformer with explicit relational encoding for math problem solving. *arXiv preprint arXiv:1910.06611*.

Jürgen Schmidhuber. 1990. Towards compositional learning in dynamic networks.

Peter Shaw, Ming-Wei Chang, Panupong Pasupat, and Kristina Toutanova. 2021. Compositional generalization and natural language variation: Can a semantic parsing approach handle both? In *Proceedings of the 59th Annual Meeting of the Association for Computational Linguistics and the 11th International Joint Conference on Natural Language Processing (Volume 1: Long Papers)*, pages 922–938.

Hongjie Shi. 2020. A sequence-to-sequence approach for numerical slot-filling dialog systems. In *Proceedings of the 21th Annual Meeting of the Special Interest Group on Discourse and Dialogue*, pages 272–277.

Satinder Pal Singh. 1992. Transfer of learning by composing solutions of elemental sequential tasks. *Machine learning*, 8(3):323–339.

Avijit Thawani, Jay Pujara, Filip Ilievski, and Pedro Szekely. 2021. Representing numbers in NLP: a survey and a vision. In *Proceedings of the 2021 Conference of the North American Chapter of the Association for Computational Linguistics: Human Language Technologies*, pages 644–656, Online. Association for Computational Linguistics.

Andrew Trask, Felix Hill, Scott E. Reed, Jack Rae, Chris Dyer, and Phil Blunsom. 2018. Neural arithmetic logic units. In *Advances in Neural Information Processing Systems*, pages 8035–8044.

Ashish Vaswani, Noam Shazeer, Niki Parmar, Jakob Uszkoreit, Llion Jones, Aidan N Gomez, Łukasz Kaiser, and Illia Polosukhin. 2017. Attention is all you need. *Advances in neural information processing systems*, 30.

Eric Wallace, Yizhong Wang, Sujian Li, Sameer Singh, and Matt Gardner. 2019. Do NLP models know numbers? Probing numeracy in embeddings. In *Proceedings of the 2019 Conference on Empirical Methods in Natural Language Processing and the 9th International Joint Conference on Natural Language Processing (EMNLP-IJCNLP)*, pages 5310–5318.

Cunxiang Wang, Boyuan Zheng, Yuchen Niu, and Yue Zhang. 2021. Exploring generalization ability of pretrained language models on arithmetic and logical reasoning. In *CCF International Conference on Natural Language Processing and Chinese Computing*, pages 758–769. Springer.

Xuezhi Wang, Jason Wei, Dale Schuurmans, Quoc Le, Ed Chi, and Denny Zhou. 2022. Self-consistency improves chain of thought reasoning in language models. *arXiv preprint arXiv:2203.11171*.

Sean Welleck, Peter West, Jize Cao, and Yejin Choi. 2021. Symbolic brittleness in sequence models: on systematic generalization in symbolic mathematics. *arXiv preprint arXiv:2109.13986*.

Eric Zelikman, Yuhuai Wu, and Noah D Goodman. 2022. Star: Bootstrapping reasoning with reasoning. *arXiv preprint arXiv:2203.14465*.

Xikun Zhang, Deepak Ramachandran, Ian Tenney, Yanai Elazar, and Dan Roth. 2020. Do language embeddings capture scales? In *Proceedings of the Third BlackboxNLP Workshop on Analyzing and Interpreting Neural Networks for NLP*, pages 292–299.

Denny Zhou, Nathanael Schärli, Le Hou, Jason Wei, Nathan Scales, Xuezhi Wang, Dale Schuurmans, Olivier Bousquet, Quoc Le, and Ed Chi. 2022. Least-to-most prompting enables complex reasoning in large language models. *arXiv preprint arXiv:2205.10625*.

Wang Zhu, Peter Shaw, Tal Linzen, and Fei Sha. 2021. Learning to generalize compositionally by transferring across semantic parsing tasks. *arXiv preprint arXiv:2111.05013*.

Yanyan Zou and Wei Lu. 2019a. Quantity tagger: A latent-variable sequence labeling approach to solving addition-subtraction word problems. In *Proceedings of the 57th Annual Meeting of the Association for Computational Linguistics*, pages 5246–5251.

Yanyan Zou and Wei Lu. 2019b. Text2Math: End-to-end parsing text into math expressions. In *Proceedings of the 2019 Conference on Empirical Methods in Natural Language Processing and the 9th International Joint Conference on Natural Language Processing (EMNLP-IJCNLP)*, pages 5330–5340.

Towards Autoformalization of Mathematics and Code Correctness: Experiments with Elementary Proofs

Garett Cunningham
School of EECS
Ohio University
Athens, OH 45701
gc974517@ohio.edu

Razvan C. Bunescu
Department of Computer Science
UNC Charlotte
Charlotte, NC 28223
razvan.bunescu@uncc.edu

David Juedes
School of EECS
Ohio University
Athens, OH 45701
juedes@ohio.edu

Abstract

The ever-growing complexity of mathematical proofs makes their manual verification by mathematicians very cognitively demanding. Autoformalization seeks to address this by translating proofs written in natural language into a formal representation that is computer-verifiable via interactive theorem provers. In this paper, we introduce a semantic parsing approach, based on the Universal Transformer architecture, that translates elementary mathematical proofs into an equivalent formalization in the language of the Coq interactive theorem prover. The same architecture is also trained to translate simple imperative code decorated with Hoare triples into formally verifiable proofs of correctness in Coq. Experiments on a limited domain of artificial and human-written proofs show that the models generalize well to intermediate lengths not seen during training and variations in natural language.

1 Introduction

To the uninitiated, the notion of mathematical proof represents simply an argument written by people to convince others of mathematical truth. However, in a real sense, mathematical proof must have formal underpinnings that go beyond the written argument. Arguments that lack such underpinnings might have fatal errors or even logical inconsistencies (see, for example, Russell's Paradox (Irvine and Deutsch, 2021)). Nevertheless, mathematical arguments written in natural language are the norm and they have great value.

In Tymoczko (1979)'s well-known paper that discusses a somewhat controversial (at the time) proof of the Four Color Theorem (Appel and Haken, 1977; Appel et al., 1977), he explores "what is a mathematical proof?" He posits that all mathematical proofs must be (i) convincing, (ii) surveyable, and (iii) formalizable. The first two points are for the reader—proofs must be convincing to and comprehensible by mathematicians. For the third point, he notes that, "Most mathematicians and philosophers believe that any acceptable proof can be formalized. We can always find an appropriate formal language and theory in which the informal proof can be embedded and 'filled out' into a rigorous formal proof." For most mathematicians, this third part is crucial for ensuring that subtle, but fatal, errors in logic do not exist in mathematical proof.

Great progress has been made since the 1970's in fully formalizing significant mathematical results. For instance, the Feit-Thompson Theorem (Gonthier et al., 2013; Gonthier, 2013) and the Four Color Theorem (Gonthier, 2008) have been formally verified using the proof assistant Coq (Bertot and Castéran, 2013), and the Kepler Conjecture (Hales, 2005; Hales et al., 2017) has been formally verified using the proof assistants Isabelle and HOL Light (Nipkow et al., 2002). Moreover, proof assistants have demonstrated immense utility for software verification, such as the full certification of a C compiler (Leroy, 2009). Proofs demonstrating the correct behavior of code share a similar structure to proofs in pure mathematics, where systems like Hoare logic replace standard first-order logic. Thus, Tymoczko's criteria for mathematical proof can be extended to the verification of programs. For many experts, LaTeX provides an excellent tool for satisfying the first two criteria. In addition, carefully written LaTeX (Higham, 2020) provides a rich structure for establishing the third criterion.

The vast majority of modern mathematics is expressed using natural language (NL), with the overwhelming majority typeset in LaTeX. Fully formalizing mathematics using proof assistants is still a difficult and time consuming task. This paper takes some preliminary steps toward bridging this gap by exploring how modern machine learning techniques can be used to convert carefully written LaTeX into equivalent, and formally verified mathematics in Coq, a process referred to as *autoformalization* in the literature (Szegedy, 2020).

Proceedings of the 1st Workshop on Mathematical Natural Language Processing (MathNLP), pages 25 - 32
December 8, 2022 ©2022 Association for Computational Linguistics

Wang et al. (2018, 2020) explored the similar task of translating mathematical statements from LaTeX into Mizar, using LSTM-based models with attention. To generate aligned LaTeX-Mizar pairs, they use a tool (Bancerek, 2006) that translates top-level Mizar statements into artificial LaTeX sentences, a task that is facilitated by the fact that Mizar is human readable and similar in length with the corresponding LaTeX version. Carman (2021) evaluated the competency of LSTMs toward formalizing a restricted set of artificially generated theorems about simple arithmetic expressions, reporting reasonable success over expression lengths seen during training. More recently, Wu et al. (2022) evaluated Codex and PaLM on a significantly more limited, but human-written set of theorems in algebra and number theory.

In contrast to prior work, we address the autoformalization of both *theorems and their proofs*, and extend the scope to *proofs of code correctness*. We use a number of manually written mathematical statements to abstract a complex grammar that is then used to generate a dataset of substantially longer and more diverse mathematical theorems and proofs. We develop an architecture based on the Universal Transformer (Dehghani et al., 2018) and adapt a copying mechanism (Gu et al., 2016) to handle arbitrary numbers and variable names at test time. The models are evaluated extensively on their ability to systematically generalize to statement lengths not seen during training, for which we report *sequence-level* accuracy as well as a *semantic-level* accuracy calculated by combining sequence-level accuracy for the theorem and running Coq to determine if the generated proof is correct. Code and data are made available at https://github.com/gc974517/autoformalization.

2 Dataset of Theorems and Proofs

We create two independent datasets of mathematical statements that overall correspond to four classes of theorems and proofs: the first dataset contains three classes of arithmetic statements (EVEN-ODD, COMPOSITES, and POWERS), described in detail in Section 2.1, and the second dataset containing statements about code correctness via Hoare logic (POLY), described in detail in Section 2.2. In each example, the input theorem-proof pair is given in LaTeX, whereas the formalized output is represented in Coq. This work focuses on the proof assistant Coq (Bertot and Castéran, 2013) because (a) there is a rich set of mathematical libraries that have been developed for it, (b) it has been used successfully to reason about significant computation artifacts, such as the ComperCert C compiler (Leroy, 2009)), and (c) it benefits from a rich set of training material for the proof assistant related to software verification (Pierce et al., 2010).

Each class of examples demonstrates features necessary for the successful autoformalization of mathematical theorems and proofs. For example, POWERS and COMPOSITES examples may define useful terminology to make the theorems shorter, e.g. proving that 4 is a *square*, or conversely they may state theorems directly without any preliminary definitions, e.g. proving $\exists n.\, n^2 = 4$. As shown in Figures 3 and 4, this corresponds in Coq to aliasing propositions using the `Definition` keyword. Additionally, the examples in the dataset provide a stress test of the copying mechanism described in Section 3.1, testing its ability to learn the correct order and number of terms to include in mathematical expressions, as well as their placement in theorems and proofs, in a way that generalizes to arbitrary tokens in mathematical language.

For each of the four classes of theorems and proofs, we manually created a few examples ourselves in order to guide the construction of a complex grammar that is then used to generate a dataset of substantially longer and more diverse mathematical theorems and proofs. Each dataset is generated using its corresponding grammar in an identical way. First, a random seed is sampled that controls the overall structure of the theorem, proof, and definition, if any. Then, the skeleton structure of the proof is completed with phrases that are sampled from a separate context-free grammar. The coarse control of the skeleton structure allows the construction of examples with interesting features like sublemmas, forward or backward proof direction, coreference, or additional conditions for the theorem, among others.

Many of the difficulties in formalizing mathematical statements from NL into Coq stem from the wide variability in the level of detail of mathematical proofs, and the frequent mismatch between what is considered an acceptable inference step in NL proofs vs. an inference step in Coq. Furthermore, there may be multiple Coq proofs for any given theorem, at different levels of granularity. We address this ambiguity by requiring the structure of the Coq proof to match the overall structure of the

LaTeX Input Sequence

Theorem. $28M + 308$ *is even.*

Proof. We know the summation between even numbers in $\mathbb{N}$ will be an even number. Observe that 308 is known to be even. Additionally, note that the pair $M \times 28$ is trivially even. This is true because the coefficient 28 is even. □

Coq Output Sequence

```
Require Import Arith.
Theorem M28_308:
  forall M : nat, Nat.even(28 * M + 308) = true.
Proof.
  intros.
  repeat rewrite Nat.even_add.
  assert (H1: Nat.even 308 = true).
  { auto. }
  assert (H2: Nat.even 28 = true).
  { auto. }
  assert (H3: Nat.even (28 * M) = true).
  { rewrite Nat.even_mul.
    rewrite H2.
    auto. }
  rewrite H1.
  rewrite H3.
  auto.
Qed.
```

Figure 1: Generated example from the EVEN-ODD set.

LaTeX Input Sequence

Theorem. $450 + a \cdot 192 + j \cdot 462$ *is guaranteed to be even for any natural terms j, and a.*

Proof. It can be justified that $192 \cdot a + j \cdot 462$ is trivially even. Note that $192a$ is an even number in $\mathbb{N}$ because multiplying between an even integer with an arbitrary number in $\mathbb{N}$ is guaranteed to be even. Likewise, $462j$ is trivially an even number in $\mathbb{N}$. The claim is proven as a consequence of the fact that the sum of even numbers with an even number will be in itself an even number. Therefore, our theorem holds. □

Coq Output Sequence

```
Require Import Arith.
Theorem a450_192j_450_even:
  forall j a : nat,
  Nat.even (192 * a + 462 * j + 450) = true.
Proof.
  intros.
  rewrite Nat.even_add.
  assert (H1: Nat.even (192 * a) = true).
  { rewrite Nat.even_mul.
    auto. }
  assert (H2: Nat.even (462 * j) = true).
  { rewrite Nat.even_mul.
    auto. }
  assert (H3: Nat.even
    (192 * a + 462 * j) = true).
  { repeat rewrite Nat.even_add.
    rewrite H1.
    rewrite H2.
    auto. }
  rewrite H3.
  auto.
Qed.
```

Figure 2: Instance of sublemma use in the EVEN-ODD dataset. The proof that the sum of non-constant terms is even (assertion H3) is given before proving the theorem.

NL proof. This is achieved by quasi-synchronously generating the LaTeX and Coq versions of mathematical statements, while still allowing for some simple re-orderings in order to improve generalization performance, e.g. swapping arguments of commutative operations.

In total, the grammar-based method for generating examples can theoretically produce over 283 million unique arithmetic examples and over 491,000 unique code examples, before considering variations in phrasing by sampling from the context-free grammar.

2.1 Arithmetic Statements

We generated three classes of mathematical statements, i.e. theorem-proof pairs:

- EVEN-ODD: an expression is even or odd.
- COMPOSITES: a number is composite.
- POWERS: a number is an integer power of n.

EVEN-ODD examples contain arithmetic expressions of n variables with even coefficients that are summed with a constant term, meaning that the parity of this constant determines the parity of the whole expression. Proofs make use of this fact with varying rigor based on our manually designed grammar, an example of which is shown by Figure 1. The Coq program is generated concurrently with the paired LaTeX example. The example shown in Figure 2 illustrates the generation and use of prior facts to prove an implicit sublemma, in both the natural language and matching Coq version.

Examples of theorems and proofs for POWERS and COMPOSITES share a similar structure in both their LaTeX and Coq forms, as shown in Figures 3 and 4, respectively. The theorems assert the existence of a natural number such that a defining property holds and their proofs are constructive, with the distinction that examples for composites prove factorization into n factors.

For both training and testing, we generate 5,000 even-odd, 5,000 composites, and 2,000 powers examples. We train on values of $n \in \{2, 3, 5, 7, 9\}$ and test on values $n \in \{2, 3, \ldots, 12\}$, where n rep-

LaTeX Input Sequence

Definition. *We define that* $w \in \mathbb{N}$ *is a* composite *natural number if taking some* R, $Q \in \mathbb{N}$ *we have* Q, $R \geq 2$ *and* $Q \times R = w$.

Theorem. 35 *is a* composite *whole number.*

Proof. Remember that a composite natural number is the multiplication between Q and R such that Q and $R \geq 2$. Allow $R = 7$, $Q = 5$. We justify the result is valid as $35 = Q \cdot R$. □

Coq Output Sequence

```
Require Import Lia.
Definition composite
  (w : nat) :=
    exists R Q : nat,
    (Q >= 2) /\ (R >= 2)
    /\ (Q * R = w).
Theorem w_composite:
  composite 35.
Proof.
  unfold composite.
  exists 7.
  exists 5.
  lia.
Qed.
```

Figure 3: Generated COMPOSITES example.

LaTeX Input Sequence

Definition. *We say* o *is a* square *whenever there exists some whole number* Z *such that* $Z \geq 2$ *and* $o = Z^2$.

Theorem. $o = 64$ *is a* square.

Proof. Let $Z = 8$. Observe that $64 = 8^2$. Also notice $Z = 8$ is more than or equal to 2. This yields 64 is a square whole number. □

Coq Output Sequence

```
Require Import Lia.
Definition square
  (o : nat) :=
    exists Z : nat,
    (Z >= 2) /\ (o = Z^2).
Theorem square_64:
  square 64.
Proof.
  unfold square.
  exists 8.
  assert (H1: 8 >= 2).
  { lia. }
  repeat split.
  apply H1.
Qed.
```

Figure 4: Generated example from the POWERS set.

resents the number of variables in the arithmetic expression, the number of factors, or the power, respectively. This is done in order to evaluate the model's ability to generalize to unseen arithmetic expression lengths and numbers of factors.

2.1.1 Handwritten Examples

We also created a small collection of 45 human-written LaTeX theorem-proof pairs to evaluate performance on examples outside of our manually generated grammar. These are distinct from the original manually written examples that were used to guide the development of the generative grammar. There are 15 examples for each type of proof from the arithmetic set, using the same vocabulary with a number of unseen grammatical structures.

2.2 Code Correctness Statements

We create a dataset of correctness proofs about short programs written in the imperative programming language *Imp* (Pierce et al., 2018), which we call POLY. The programs represent various algorithms for evaluating a polynomial, and their proofs of correctness verify that the programs correctly model the polynomial as a mathematical function. Proofs are conducted as either fully decorated programs or as sequences of Hoare triples with natural language justifying steps in between. An example is shown in Figure 5.

For both training and testing data, we generate 5,000 examples. We train on programs containing 2, 3, 5, 7, 9, and 11 lines, then test on programs containing from 2 up to 14 lines to evaluate the model's ability to generalize to novel program lengths.

3 Semantic Parsing Architecture

To formalize LaTeX statements into Coq, we developed an encoder-decoder architecture based on the Universal Transformer (Dehghani et al., 2018). Similar to Csordás et al. (2021), we do so by adding recursive passes into the encoder and decoder of a base Transformer (Vaswani et al., 2017), thus making the model analogous to a Universal Transformer without adaptive computation time (ACT). Further, we introduce a copying mechanism and support for out-of-vocabulary mathematical terms.

3.1 Copying Mechanism

Mathematical language contains features uncommon or non-existent in natural language, such as numbers, variables, and carefully defined terminology. In order to address the use of general mathematical jargon, these tokens are replaced in the LaTeX input with generic forms denoting their usage, such as `<var1>` up to `<varN>` for variables,

LaTeX Input Sequence	Coq Output Sequence

Theorem. *Consider the following series of commands such that*

```
S := 3;
S := 3 + S * Z;
S := 1 + S * Z
```

Allow $Z = y$*, for any natural number* y*, ahead of running this code then* $S = 3 \times y^2 + 3 \times y + 1$ *after the set of instructions has executed.*

Proof. By application of usual Hoare logic:

$$\{Z = y\}$$

`S := 3;`

$$\{Z = y \land S = 3\}$$

`S := 3 + S * Z;`

$$\{Z = y \land S = 3 \times y + 3\}$$

`S := 1 + S * Z`

$$\{Z = y \land S = 3 \times y^2 + 3 \times y + 1\}$$

Hence, this program is shown to be correct. □

```
Require Import String.
From PLF Require Import Imp.
From PLF Require Import Hoare.
Theorem poly_code_correct:
  forall y : nat,
  {{ Z = y }}
  S := 3;
  S := 3 + S * Z;
  S := 1 + S * Z
  {{ S = 3 * y ^ 2 + 3 * y + 1 }}.
Proof.
  intros.
  apply hoare_seq with
    (Q := (
      (Z = y /\ S = 3)
    )%assertion).
  apply hoare_seq with
    (Q := (
      (Z = y /\ S = 3 * y + 3)
    )%assertion).
  apply hoare_seq with
    (Q := (
      (Z = y /\ S = 3 * y^2 + 3 * y + 1)
    )%assertion).
  all: eapply hoare_consequence_pre;
  try (apply hoare_asgn || assn_auto'').
Qed.
```

Figure 5: Generated POLY example: [Left] the Hoare logic proof; [Right] the code correctness proof in Coq.

which effectively ensures *generalization to variable renaming* (Ferreira et al., 2022), `<nat1>` up to `<natN>` for numbers, or `<def>` for definitions, coupled with the use of a copying mechanism adapted from Gu et al. (2016). Note that a different generic token is introduced for each unique numerical constant or variable literal in the theorem and its proof, and the corresponding generic token is used in the Coq version. For example, considering the ⟨LaTeX, Coq⟩ pair in Figure 3, `<nat1>`, `<nat2>`, `<nat3>`, and `<nat4>` would be used to replace the constants 2, 35, 7, and 5 respectively, everywhere in the LaTeX and Coq statements. Similarly, `<var1>`, `<var2>`, and `<var3>` were used to replace variable literals w, R, and Q. This is in contrast to using just two generic tokens `<nat>` and `<var>` everywhere, which would make all numbers coreferent and all variables coreferent. Preliminary experiments validated the utility of encoding these distinctions while maintaining the correct coreference in both LaTeX and Coq statements.

Overall, by using generic tokens for numbers, variables, and definitions, only a limited set of embeddings need to be trained and the model is forced to utilize contextual information in order to appropriately copy tokens into the Coq output. In this way, the model has the ability to generalize to unseen numbers or variable and definition names.

The original CopyNet (Gu et al., 2016) used an encoder-decoder architecture with a copying mechanism to calculate the probabilities of generating in-vocabulary tokens vs. copying tokens from the input sequence to the output. Our autoformalization task guarantees mutual exclusivity between generating (g) and copying (c) tokens, which allows using a simplified formula for calculating the probability of producing a token y_t at time step t. Letting $\mathcal{V}_c$ denote the Coq vocabulary, X denote the input sequence of LaTeX tokens, and $\mathcal{X}$ denote the collection of unique tokens in X, we calculate the probability of producing y_t as:

$$p(y_t) = \begin{cases} p(y_t, \mathrm{g}) = \dfrac{1}{Z_t} e^{\psi_g(y_t)}, & y_t \in \mathcal{V}_c \\ p(y_t, \mathrm{c}) = \dfrac{1}{Z_t} \displaystyle\sum_{x_j \in X : x_j = y_t} e^{\psi_c(x_j)}, & y_t \in \mathcal{X} \end{cases}$$

where $Z_t = \sum_{y_t \in \mathcal{V}_c} e^{\psi_g(y_t)} + \sum_{x_j \in X} e^{\psi_c(x_j)}$. The scoring functions are given by $\psi_g(y_t) = \mathbf{v}_{y_t}^\top \mathbf{W}_o \mathbf{s}_t$ and $\psi_c(x_j) = \tanh\left(\mathbf{h}_j^\top \mathbf{W}_c\right)\mathbf{s}_t$, where $\mathbf{v}_{y_t}$ is a one-hot encoding of y_t, $\mathbf{h}_j$ is the hidden encoder state for the input token x_j, $\mathbf{s}_t$ is the decoder state at step t, and $\mathbf{W}_o$ and $\mathbf{W}_c$ are learnable parameters.

3.2 Encoder-Decoder Architecture

We diverge from the standard Transformer architecture in a few crucial ways:

- Probabilities are calculated via $p(y_t)$ above.
- Absolute positional encodings are removed.
- Self-attention uses relative positional representations as in Shaw et al. (2018).
- Stacks of N encoder/decoder blocks have T recurrent passes.

All other aspects of the model remain unchanged from the original Transformer. We emphasize relative positional information over absolute in our model architecture. Preliminary evaluations on the EVEN-ODD dataset showed that Transformer models that use absolute positional encodings obtain 0% sequence-level accuracy on expression lengths that are not seen at training time. Removing reliance on absolute position resolves this type of systematic generalization. The use of relative positional encodings for the Transformer-based models was thus essential for achieving stronger systematic generalization, which also agrees with the findings of Csordás et al. (2021) on other NLP tasks.

4 Experimental Evaluations

To evaluate the performance of trained models, we ran two primary experiments: first on the collection of arithmetic examples, then on the collection of code correctness examples. All models are evaluated in terms of *sequence-level* accuracy, where an example is considered correctly processed only if the generated Coq sequence for both the theorem and its proof perfectly matches token by token the ground truth sequence. We also report *semantic-level* accuracy, for which the generated Coq theorem needs to attains a perfect sequence-level accuracy and the Coq engine verifies that the generated Coq proof truly proves the generated Coq theorm, regardless of whether it matches the ground truth version of the proof. This emphasizes that the model was able to capture the general meaning of the natural language proof by correctly translating the theorem and successfully proving it using the natural language version as a guide.

All experiments were performed on one NVIDIA RTX-A6000 GPU with 48GB of memory.

	EVEN-ODD		COMPOSITES		POLY
n	Seq	Sem	Seq	Sem	Both
2	99.6	99.8	76.7	97.6	100.0
3	99.4	99.6	64.6	94.2	100.0
4	99.4	99.4	56.1	93.9	82.1
5	99.2	99.6	54.9	94.4	99.2
6	98.8	98.8	57.1	94.3	45.1
7	99.1	99.5	58.5	93.4	96.5
8	93.8	94.0	53.5	88.3	15.7
9	98.6	98.6	53.7	93.7	98.2
10	7.0	7.0	1.2	1.6	35.6
11	0.0	0.0	0.0	0.0	93.5
12+	0.0	0.0	0.0	0.0	0.0

POWERS	Seq = 100.0	Sem = 100

Table 1: Sequence-level (Seq) and semantic-level (Sem) accuracy (%) on test examples, split by expression length, with the exception of POWERS.

4.1 Arithmetic Statements

We evaluate a Transformer model on the full data combining EVEN-ODD + COMPOSITES + POWERS and using both the theorem and its proof in each sequence. We tune a model with embedding and state sizes of 32, a feed forward width of 256, 4 encoder and decoder blocks with 4 recurrent passes, 4 attention heads, and a clipping value of 2 for self-attention. We trained this model over minibatches of size 20, optimized with Adam using $\beta_1 = 0.9$, $\beta_2 = 0.98$, $\varepsilon = 1e - 9$, and an initial learning rate of 0.001, annealed by a factor of $1/\sqrt{10}$ based on training loss plateaus with a patience of 5 epochs.

The results in Table 1 show that the model generalizes well to the intermediate lengths of $\{4, 6, 8\}$, with a small number of correctly translated examples longer than the maximum of 9 used in training. Otherwise, the model fails to generalize to longer unseen lengths, which is not surprising, given that Transformer models are known to fail dramatically at systematic generalization on longer inputs for various NLP tasks (Csordás et al., 2021), or to incur substantial decrease in accuracy for longer symbolic integration problems (Welleck et al., 2022). Switching to semantic-level evaluation leads to a significant increase in accuracy for COMPOSITES, with a more modest increase for EVEN-ODD.

4.2 Code Correctness Statements

We extend our scope to include data representing proofs of program correctness using the language of Hoare logic. We train a separate model with the same embedding and state sizes, feed forward width, and learning rates as in Section 4.1. Depth is increased to 8 encoder and decoder blocks with 8 recurrent passes, 8 attention heads, and a clipping value of 8. The model is trained over minibatches of size 1 with Adam, with a patience of 3 epochs.

The POLY results shown in Table 1 demonstrate that the model is able to generalize to program line counts of $\{4, 6, 8, 10\}$ unseen during training with diminishing returns as the program length grows, eventually failing to generalize for lengths longer than the maximum seen in training. We observe that increasing the depth of the model significantly improved generalization. A model with identical hyperparameters to the arithmetic experiment yielded less then half the sequence-level accuracy for intermediate program lengths. Therefore, further increasing the depth of the model could push performance closer to optimal generalization to intermediate lengths at the cost of significantly more computing resources. Additionally, POLY examples are far less prone to non-fatal token swapping errors. We observe that semantic-level accuracy is identical to sequence-level, as all copying errors compromised the validity of the proof. Therefore, accuracies are shown as one column (Both).

4.3 Handwritten Examples

We also evaluate the semantic-level accuracy of the trained models on the collection of 45 human-written LaTeX theorem-proof pairs. This is done by manually verifying that the generated Coq theorem corresponds to the LaTeX version and that the subsequent proof is correct according to the Coq interpreter. The fully trained model achieved 53.3% for both EVEN-ODD and COMPOSITES, and 73.3% for POWERS.

Mistakes in almost all cases are confined to the mishandling of out-of-vocabulary tokens, such as mis-copying a variable within a definition or the omission of an assertion in the proof tied to a term. The model otherwise generated syntactically sound Coq code. Mistakes strongly correlate with examples that deviate significantly from the grammatical structure of the artificial data. Thus, pre-trained language models as evaluated by Wu et al. (2022) or pre-training new models on mathematical corpora like MATH (Hendrycks et al., 2021) may serve to alleviate the problems caused by the scarcity of aligned natural and formal mathematics data.

5 Concluding Remarks

As we have seen, it is feasible to train machine learning models to perform autoformalization over very restricted domains of math and code correctness proofs. These models show capability to systematically generalize to new expression lengths and program sizes. Moreover, these models were able to translate previously unseen hand written natural language examples, albeit with lower accuracy. We are hopeful that this approach can be applied to autoformalization of a larger segment of mathematics and code verification.

As mentioned by Szegedy (2020), "Autoformalization is not just a challenge: successful autoformalization would represent a breakthrough for general AI with significant implications in various domains." We see an especially significant impact in education, where integration of autoformalization into proof assistants for introductory mathematics and software verification courses would enable the detection of missing steps or misconceptions in students' proofs.

References

K. Appel and W. Haken. 1977. Every planar map is four colorable, part I: discharging. *Illinois Journal of Mathematics*, 21(3):429 – 490.

K. Appel, W. Haken, and J. Koch. 1977. Every planar map is four colorable. II: Reducibility. *Ill. J. Math.*, 21:491–567.

Grzegorz Bancerek. 2006. Automatic translation in formalized mathematics. *Mechanized Mathematics and Its Applications*, 5(2):19–31.

Yves Bertot and Pierre Castéran. 2013. *Interactive theorem proving and program development: Coq'Art: the calculus of inductive constructions*. Springer Science & Business Media.

Benjamin Andrew Carman. 2021. *Translating LaTeX to Coq: A Recurrent Neural Network Approach to Formalizing Natural Language Proofs*. Ph.D. thesis, Ohio University.

Róbert Csordás, Kazuki Irie, and Juergen Schmidhuber. 2021. The devil is in the detail: Simple tricks improve systematic generalization of transformers. In *Proceedings of the 2021 Conference on Empirical Methods in Natural Language Processing*, pages 619–634.

Mostafa Dehghani, Stephan Gouws, Oriol Vinyals, Jakob Uszkoreit, and Łukasz Kaiser. 2018. Universal transformers. *arXiv preprint arXiv:1807.03819*.

Deborah Ferreira, Mokanarangan Thayaparan, Marco Valentino, Julia Rozanova, and Andre Freitas. 2022. To be or not to be an Integer? Encoding Variables for Mathematical Text. In *Findings of the Association for Computational Linguistics: ACL 2022*, pages 938–948, Dublin, Ireland. Association for Computational Linguistics.

G. Gonthier. 2013. Engineering mathematics: The odd order theorem proof. In *Conference Record of the Annual ACM Symposium on Principles of Programming Languages*, pages 1–2. Cited By 12.

Georges Gonthier. 2008. Formal proof – the four-color theorem. *Notices of the AMS*, 55(11):1382–1393.

Georges Gonthier, Andrea Asperti, Jeremy Avigad, Yves Bertot, Cyril Cohen, François Garillot, Stéphane Le Roux, Assia Mahboubi, Russell O'Connor, Sidi Ould Biha, Ioana Pasca, Laurence Rideau, Alexey Solovyev, Enrico Tassi, and Laurent Théry. 2013. A machine-checked proof of the odd order theorem. In *Interactive Theorem Proving*, pages 163–179, Berlin, Heidelberg. Springer Berlin Heidelberg.

Jiatao Gu, Zhengdong Lu, Hang Li, and Victor OK Li. 2016. Incorporating copying mechanism in sequence-to-sequence learning. In *Proceedings of the 54th Annual Meeting of the Association for Computational Linguistics (Volume 1: Long Papers)*, pages 1631–1640.

Thomas Hales, Mark Adams, Gertrud Bauer, Tat Dat Dang, John Harrison, Le Truong Hoang, Cezary Kaliszyk, Victor Magron, Sean McLaughlin, Tat Thang Nguyen, and et al. 2017. A formal proof of the Kepler conjecture. *Forum of Mathematics, Pi*, 5:e2.

Thomas C. Hales. 2005. A proof of the Kepler conjecture. *Ann. Math. (2)*, 162(3):1065–1185.

Dan Hendrycks, Collin Burns, Saurav Kadavath, Akul Arora, Steven Basart, Eric Tang, Dawn Song, and Jacob Steinhardt. 2021. Measuring mathematical problem solving with the math dataset. *NeurIPS*.

Nicholas J. Higham. 2020. *Handbook of Writing for the Mathematical Sciences*, third edition. Society for Industrial and Applied Mathematics, Philadelphia, PA, USA.

Andrew David Irvine and Harry Deutsch. 2021. Russell's Paradox. In Edward N. Zalta, editor, *The Stanford Encyclopedia of Philosophy*, Spring 2021 edition. Metaphysics Research Lab, Stanford University.

Xavier Leroy. 2009. Formal verification of a realistic compiler. *Communications of the ACM*, 52(7):107–115.

Tobias Nipkow, Lawrence C Paulson, and Markus Wenzel. 2002. *Isabelle/HOL: a proof assistant for higher-order logic*, volume 2283. Springer Science & Business Media.

Benjamin C Pierce, Chris Casinghino, Marco Gaboardi, Michael Greenberg, Cătălin Hriţcu, Vilhelm Sjöberg, and Brent Yorgey. 2010. Software foundations. *Webpage: http://www. cis. upenn. edu/bcpierce/sf/current/index. html.*

Benjamin C. Pierce, Arthur Azevedo de Amorim, Chris Casinghino, Marco Gaboardi, Michael Greenberg, Cătălin Hriţcu, Vilhelm Sjöberg, Andrew Tolmach, and Brent Yorgey. 2018. *Programming Language Foundations*. Software Foundations series, volume 2. Electronic textbook. Version 5.5. `http://www.cis.upenn.edu/~bcpierce/sf`.

Peter Shaw, Jakob Uszkoreit, and Ashish Vaswani. 2018. Self-attention with relative position representations. In *Proceedings of the 2018 Conference of the North American Chapter of the Association for Computational Linguistics: Human Language Technologies, Volume 2 (Short Papers)*, pages 464–468, New Orleans, Louisiana. Association for Computational Linguistics.

Christian Szegedy. 2020. A promising path towards autoformalization and general artificial intelligence. In *Intelligent Computer Mathematics. CICM 2020. Lecture Notes in Computer Science*, volume 12236, pages 3–20. Springer.

Thomas Tymoczko. 1979. The four-color problem and its philosophical significance. *Journal of Philosophy*, 76(2):57–83.

Ashish Vaswani, Noam Shazeer, Niki Parmar, Jakob Uszkoreit, Llion Jones, Aidan N Gomez, Łukasz Kaiser, and Illia Polosukhin. 2017. Attention is all you need. *Advances in Neural Information Processing Systems*, 30:5998–6008.

Qingxiang Wang, Chad Brown, Cezary Kaliszyk, and Josef Urban. 2020. Exploration of neural machine translation in autoformalization of mathematics in Mizar. In *Proceedings of the 9th ACM SIGPLAN International Conference on Certified Programs and Proofs*, CPP 2020, page 85–98, New York, NY, USA. Association for Computing Machinery.

Qingxiang Wang, Cezary Kaliszyk, and Josef Urban. 2018. First experiments with neural translation of informal to formal mathematics. In *Intelligent Computer Mathematics*, pages 255–270, Cham. Springer International Publishing.

Sean Welleck, Peter West, Jize Cao, and Yejin Choi. 2022. Symbolic Brittleness in Sequence Models: On Systematic Generalization in Symbolic Mathematics. *Proceedings of the AAAI Conference on Artificial Intelligence*, 36(8):8629–8637. Number: 8.

Yuhuai Wu, Albert Q Jiang, Wenda Li, Markus N Rabe, Charles Staats, Mateja Jamnik, and Christian Szegedy. 2022. Autoformalization with large language models. *arXiv preprint arXiv:2205.12615*.

Numerical Correlation in Text

Daniel Spokoyny
Carnegie Mellon University

Chien-Sheng Wu
Salesforce AI Research

Caiming Xiong,
Salesforce AI Research

Abstract

Evaluation of quantitative reasoning of large language models is an important step towards understanding their current capabilities and limitations. We propose a new task, Numerical Correlation in Text, which requires models to identify the correlation between two numbers in a sentence. To this end, we introduce a new dataset, which contains over 2,000 Wikipedia sentences with two numbers and their correlation labels. Using this dataset we are able to show that recent numerically aware pretraining methods for language models do not help generalization on this task posing a challenge for future work in this area.[1]

1 Introduction

Numerical reasoning tasks are one area where the performance of Large Language Models (LLMs) has not improved as drastically (Rae et al., 2021) as on other tasks. Good performance is critical for many downstream applications in areas such as fact checking, question-answering, or search. Different tasks have been proposed to evaluate the numerical reasoning capabilities of LLMs (Mishra et al., 2022).

We can analyze these tasks along two dimensions: diversity of knowledge required and how solvable the task is. Higher diversity ensures better coverage across different domains while higher solvability yields more interpretable metrics. Mathematical word problems (MWPs) are written in a way that the text of the problem is always sufficient to determine the exact unique answer and are therefore highly solvable. However, they lack in diversity since many MWP datasets are constructed from templates or are even fully synthetic.

In contrast, numerical cloze-style problems requires highly diverse knowledge since they can be easily formed from any text that includes numbers.

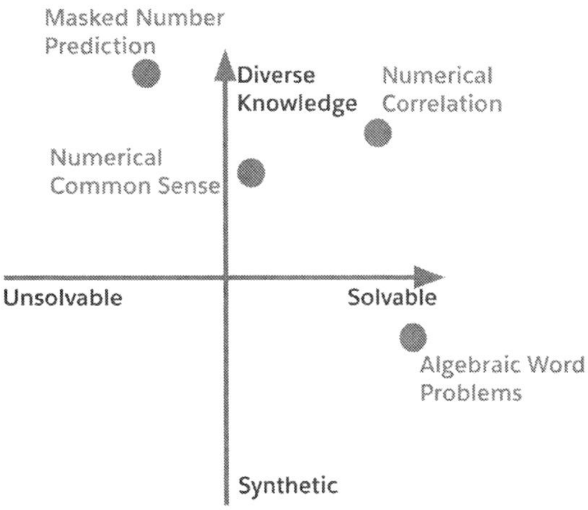

Figure 1: An illustrative plot of certain numerical evaluation tasks along the two dimensions of diversity and solvability. Our aim with numerical correlation is for the task to be both diverse and solvable.

A consequence of formulating cloze-style problems is that many texts do not provide sufficient information to determine the correct answer and have inherent uncertainty which results in a lower solvability. As an example from the NumerSense dataset (Lin et al., 2020), "Some plant varieties can grow up to <mask> feet tall." In Figure 1, we show an illustrative plot of tasks along these two dimensions. A good numeracy evaluation task should be both diverse and solvable.

In this work we propose Numerical Correlation in text, a new task that aims to retain both high diversity and high solvability. Given two numbers in text the task is to predict whether the numbers are positively, negatively or not correlated. For example: "Some plant varieties can grow up to 6 feet tall and require 20 liters of water a month". We expect a positive correlation between the height of the plant and the amount of water it would need. This shows the key insight that predicting the correlation relationship between two numbers is possible with-

[1]Work completed during internship at Salesforce Research. Please direct correspondence to: dspokoyn@cs.cmu.edu

Proceedings of the 1st Workshop on Mathematical Natural Language Processing (MathNLP), pages 33 - 39
December 8, 2022 ©2022 Association for Computational Linguistics

# Ex	Text	Label
1.	The president travels on average **30** times a year on Air Force one a Boeing **747**.	Neutral
2.	A **2** bedroom, **1800** square feet house is hard to find in this neighborhood.	Positive
3.	To cook a 20 lb turkey place in the oven for **2** hours at **435** degrees.	Negative

Table 1: Explanations for the three examples: 1) the model of the plane should not change how often the president travels, 2) we expect more bedrooms to increase the size of the house, and 3) we expect an increase of temperature to decrease the cooking time.

out having to exactly predict the missing numbers. The task of numerical correlation requires a variety of commonsense reasoning skills but is trained with a cross-entropy objective and evaluated with a simple accuracy metric. We provide examples of sentences and their labels in Table 1.

Although correlation between two numbers can involve incredibly complex functions, we approximate the correlation to be linear and treat it as a three-way classification. We use a qualification task to select a group of Amazon Mechanical Turk (AMT) labelers and construct a dataset of Wikipedia sentences which contain two numbers and their correlation relationship.

We investigate the performance of four models: two general pretrained language transformers and two numerically aware models on our new dataset in a few-shot setting. When probed on the numerical correlation task we see that all models exhibit a plateau in their performance with only 6% of the training data. Further all models underperform the human baseline in both the finetuning and linear probing setting. Surprisingly, our results also indicate that existing numerically pretraining methods do not result in better performance on the numerical correlation task.

2 Dataset

2.1 Qualification

We used ten handwritten numerical correlation examples and had 100 AMT workers with 99%> approval rate label them. On average each question took around 1 minute to complete. Thresholding on 80% accuracy or above left us with 18 AMTlabelers. Examples and the instructions are shown in the Appendix Table 2 and Figure 5, respectively.

2.2 Annotation

We use the WikiConvert dataset (Thawani et al., 2021) which contains over 900k sentences with at least one measurement in each sentence. We use the three original correlation labels (Positive, Negative, Neutral)[2] and had each sentence labeled by three different AMTlabelers. We selected 1,000 random sentences that contain two measurements and another 1,000 sentences that contain two any two quantities.[3]

We used Krippendorff alpha to measure the inter-annotator agreement and found that the agreement was 0.55 (scale is [-1,1]). We computed an average "Jackknife" F1 score of 77 by choosing one label to be the ground truth and averaging the F1 score of the other two labels. We also observe that the time taken to label each sentence rose to 1.7 minutes on average, likely due to the increased difficulty to ascertain the correlation in random sentences.

2.2.1 Negative

Out of the 2,000 sentences only 42 were found to have a negative correlation which is too few data points to train or evaluate a model. For this reason we experimented with two strategies to generate more negative correlation examples: 1) editing a measurement in real sentence 2) providing a description of a real negative relationship and prompting labelers to provide a sentence as an example. In a small pilot we found that the first strategy was incredibly more time consuming to complete and so we only used the second strategy to generate negative correlation examples. We provided 60 descriptions of negative relationships and asked the three labelers to provide an example for each sentence.[4] In total our dataset consists of 124 sentences with negative correlation, 746 with positive correlation and 1,155 with neutral correlation.

[2]We introduce a fourth label (Unanswerable) which we advised the labelers to use sparingly when they were unsure of the answer

[3]We filtered out sentences that contained dates or where shorter than 64 characters in length.

[4]We hand filtered out sentences that did not properly follow the instructions.

	Test F1 (Neutral, Positive, Negative)				
w/ 10% Train	GenBERT	GeMM	RoBERTa-Base	Bart-Large	Human Jackknife
Linear Probing	33.0 (71.1 / 26.7 / 0.1)	37.9 (72.3 / 41.6 / 0)	23.7 (71.1 / 0 / 0)	**64.9 (76.6 / 57.7 / 60.4)**	~77
Finetuning	62.1 (77.5 / 59.3 / 49.6)	66.7 (77.9 / 65.5 / 56.8)	**69.6 (80.7 / 66.3 / 61.8)**	68.6 (* / * / *)	

Figure 2: Summary of the performance of the four models on the numerical correlation task with 10% of the training data.

3 Experiments

Given a sentence X and two numbers y_1 and y_2 in the text, we define the task of predicting the correlation between the two numbers as a classification task with the label set $C = \{Positive, Negative, Neutral\}$. We compare four models, two general pretrained language models (BART (Lewis et al., 2020) and RoBERTa (Liu et al., 2019)) and two numerically aware models (GeMM (Spokoyny et al., 2022) and GenBERT (Geva et al., 2020)). We conduct few-shot learning experiments where the model is trained on between 1% to 10% of the training data and the remaining data is split into a validation and test set evenly. We train all models with the AdamW optimizer (Loshchilov and Hutter, 2019) with a learning rate of 1e-5 and a batch size of 16. We use the majority vote labeling to choose the final label for each sentence in all subsequent experiments.[5] We report the test F1 scores averaged over 5 initialization seeds.

3.1 Supervised

We conduct few-shot linear probing as well as full finetuning experiments and plot the results in Figure 3 and Figure 4 respectively. For our linear probing experiments we freeze the parameters of the model and only train a linear classifier, $W_\theta \in \mathbb{R}^{d\times 3}$, where d is the hidden size of the model. We observed that BART performed better by a large margin (20 F1) as compared to the second best performing model, GeMM. However, all models experience a plateau in performance after only 6% of the training data.

Unlike the linear probing experiments, when we finetune the models we observe that all models (except GenBERT) converge to similar performance,

[5]In case of a tie we do not use the sentence in our data.

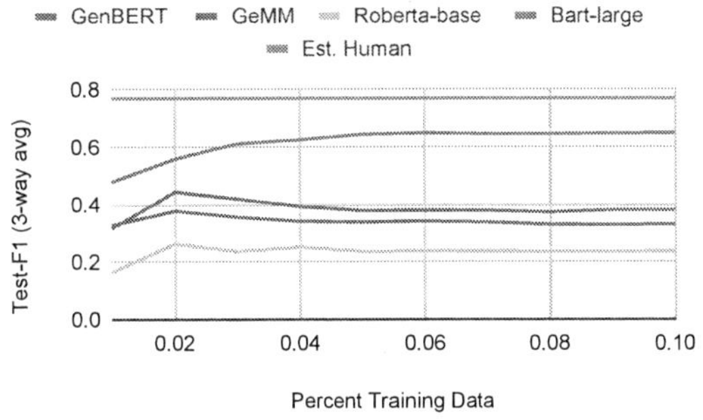

Figure 3: Linear probing experiments with 1% to 10% of the training data.

approximately 10 F1 points below human performance. The poor performance of GenBERT could be explained by the fact that it uses a BERT architecture whilst the other models are based on RoBERTa and BART. We present all of the supervised Test-F1 results with 10% of the training data in Figure 2.

3.2 Unsupervised

Since we observe the actual values of the both numbers we can probe a model in an unsupervised fashion to predict the correlation relationship. We do this by selecting one number (y_1) to be the target prediction and masking it's value in the sentence. We then probe the model to predict the value of the target (y_1) with different values of the other number (y_2). We use GeMM, a numerically pretrained model (Spokoyny et al., 2022) and denominate the model's prediction for the masked value as $\hat{Y}$.

We construct $\mathcal{N}$ examples, $\{X_1, X_\mathcal{N}\}$, by selecting values linearly spaced between $\{y_2{*}0.5, y_2{*}2\}$ and pass each example to the model to predict the $\mathcal{N}$ values of $\{\hat{Y}_1, \hat{Y}_\mathcal{N}\}$. We can then calculate the R-squared values of the linear regression for each

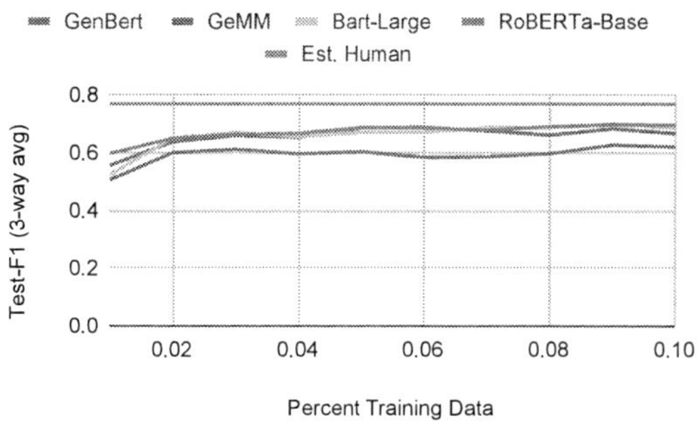

Figure 4: Full finetuning experiments with 1% to 10% of the training data.

pair of numbers in a sentence. We pick a threshold value τ and build a deterministic classifier which predicts "Neutral" if the R-squared value is less than τ, "Positive" if the R-squared value is greater than τ and the slope is positive, and "Negative" if the R-squared value is greater than τ and the slope is negative. When evaluated on a held out test set this classifier performs close to randomly guessing the label.

4 Related Work

4.1 Numerical Reasoning

An active area of research in NLP is focused on solving numerical reasoning tasks. There have been many datasets collected such as AQuA-RAT (Ling et al., 2017), Dolphin18K (Huang et al., 2016), Math23K (Wang et al., 2017), MathQA (Amini et al., 2019) which contain a mathematical question expressed in natural language and an answer. Benchmarks which aim to evaluate the general abilities of LLMs like BIG-bench, have also incorporated numerical reasoning tasks such as arithmetic questions or unit conversion (Srivastava and et al., 2022). To solve these problems a model needs to perform certain necessary calculations to arrive at the answer. Typically the value of the numbers provide no information to help disambiguate the derivation of the solution and can be treated symbolically. One key aspect of these tasks is that there exists no ambiguity in the answer.

4.2 Commonsense Reasoning

Another area of research has focused on cloze-style prediction of numbers in textual contexts. Certain works have limited the output space of numbers to small ranges (Lin et al., 2020), their exponent value (Chen et al., 2019) whilst others have aimed to produce distributions over the entire real number line (Spithourakis and Riedel, 2018; Spokoyny and Berg-Kirkpatrick, 2020). As opposed to the previous section, these tasks commonly do not have a correct answer but are ambiguous. A great advantage of numerical cloze-style reasoning is the ubiquity of available data in different forms and domains. However, it is difficult to measure progress and interpret the evaluation metrics such as likelihood for these types of commonsense tasks.

There are other NLP tasks which have concentrated on the difficulties that arise when numbers are present in a text. Ravichander et al. (2019) proposed EQUATE, a benchmark quantitative reasoning in natural language inference while other works have focused on quantity entailment (Roy et al., 2015). Dubey et al. (2019) built a dataset where the numerical values were useful to predict the sentiment of sarcastic tweets. Sundararaman et al. (2022) proposed a classification task of numbers into entities (Count, Size, Year, Percentage, Date, Age), while similar work has considered the problem of solving numeric Fused-Heads (Elazar and Goldberg, 2019). Our work on the correlation task focuses on a particular relationship between two quantities in text. However there are others potential relationships between numbers in text that could be explored such as causation.

5 Conclusion

We introduced a new task of predicting numerical correlation in text and build an annotated dataset to evaluate models on this task. Using this dataset we show that pretrained language models have poor performance on this task and that current methods to add numerically aware pretraining to models are not effective. We identified that there exists a large gap between human performance and the best supervised model. In the future we hope to expand our annotation to include the slope of the correlation. We believe that predicting both the slope and correlation type of two numbers can be improve interpretability in numerical question answering and commonsense reasoning applications. In future work we also plan to expand the dataset to capture numerical correlation relationships in longer chunks of text such as paragraphs and documents.

References

Aida Amini, Saadia Gabriel, Shanchuan Lin, Rik Koncel-Kedziorski, Yejin Choi, and Hannaneh Hajishirzi. 2019. MathQA: Towards interpretable math word problem solving with operation-based formalisms. In *Proceedings of the 2019 Conference of the North American Chapter of the Association for Computational Linguistics: Human Language Technologies, Volume 1 (Long and Short Papers)*, pages 2357–2367, Minneapolis, Minnesota. Association for Computational Linguistics.

Chung-Chi Chen, Hen-Hsen Huang, Hiroya Takamura, and Hsin-Hsi Chen. 2019. Numeracy-600K: Learning numeracy for detecting exaggerated information in market comments. In *Proceedings of the 57th Annual Meeting of the Association for Computational Linguistics*, pages 6307–6313, Florence, Italy. Association for Computational Linguistics.

Abhijeet Dubey, Lakshya Kumar, Arpan Somani, Aditya Joshi, and Pushpak Bhattacharyya. 2019. "when numbers matter!": Detecting sarcasm in numerical portions of text. In *Proceedings of the Tenth Workshop on Computational Approaches to Subjectivity, Sentiment and Social Media Analysis*, pages 72–80, Minneapolis, USA. Association for Computational Linguistics.

Yanai Elazar and Yoav Goldberg. 2019. Where's my head? Definition, data set, and models for numeric fused-head identification and resolution. *Transactions of the Association for Computational Linguistics*, 7:519–535.

Mor Geva, Ankit Gupta, and Jonathan Berant. 2020. Injecting numerical reasoning skills into language models. In *Proceedings of the 58th Annual Meeting of the Association for Computational Linguistics*, pages 946–958, Online. Association for Computational Linguistics.

Danqing Huang, Shuming Shi, Chin-Yew Lin, Jian Yin, and Wei-Ying Ma. 2016. How well do computers solve math word problems? large-scale dataset construction and evaluation. In *Proceedings of the 54th Annual Meeting of the Association for Computational Linguistics (Volume 1: Long Papers)*, pages 887–896, Berlin, Germany. Association for Computational Linguistics.

Mike Lewis, Yinhan Liu, Naman Goyal, Marjan Ghazvininejad, Abdelrahman Mohamed, Omer Levy, Veselin Stoyanov, and Luke Zettlemoyer. 2020. BART: Denoising sequence-to-sequence pre-training for natural language generation, translation, and comprehension. In *Proceedings of the 58th Annual Meeting of the Association for Computational Linguistics*, pages 7871–7880, Online. Association for Computational Linguistics.

Bill Yuchen Lin, Seyeon Lee, Rahul Khanna, and Xiang Ren. 2020. Birds have four legs?! numersense: Probing numerical commonsense knowledge of pretrained language models. *ArXiv*, abs/2005.00683.

Wang Ling, Dani Yogatama, Chris Dyer, and Phil Blunsom. 2017. Program induction by rationale generation: Learning to solve and explain algebraic word problems. In *Proceedings of the 55th Annual Meeting of the Association for Computational Linguistics (Volume 1: Long Papers)*, pages 158–167, Vancouver, Canada. Association for Computational Linguistics.

Yinhan Liu, Myle Ott, Naman Goyal, Jingfei Du, Mandar Joshi, Danqi Chen, Omer Levy, Mike Lewis, Luke Zettlemoyer, and Veselin Stoyanov. 2019. Roberta: A robustly optimized bert pretraining approach. *ArXiv*, abs/1907.11692.

Ilya Loshchilov and Frank Hutter. 2019. Decoupled weight decay regularization. In *ICLR*.

Swaroop Mishra, Arindam Mitra, Neeraj Varshney, Bhavdeep Sachdeva, Peter Clark, Chitta Baral, and Ashwin Kalyan. 2022. NumGLUE: A suite of fundamental yet challenging mathematical reasoning tasks. In *Proceedings of the 60th Annual Meeting of the Association for Computational Linguistics (Volume 1: Long Papers)*, pages 3505–3523, Dublin, Ireland. Association for Computational Linguistics.

Jack W. Rae, Sebastian Borgeaud, Trevor Cai, Katie Millican, Jordan Hoffmann, Francis Song, John Aslanides, Sarah Henderson, Roman Ring, Susannah Young, Eliza Rutherford, Tom Hennigan, Jacob Menick, Albin Cassirer, Richard Powell, George van den Driessche, Lisa Anne Hendricks, Maribeth Rauh, Po-Sen Huang, Amelia Glaese, Johannes Welbl, Sumanth Dathathri, Saffron Huang, Jonathan Uesato, John F. J. Mellor, Irina Higgins, Antonia Creswell, Nathan McAleese, Amy Wu, Erich Elsen, Siddhant M. Jayakumar, Elena Buchatskaya, David Budden, Esme Sutherland, Karen Simonyan, Michela Paganini, L. Sifre, Lena Martens, Xiang Lorraine Li, Adhiguna Kuncoro, Aida Nematzadeh, Elena Gribovskaya, Domenic Donato, Angeliki Lazaridou, Arthur Mensch, Jean-Baptiste Lespiau, Maria Tsimpoukelli, N. K. Grigorev, Doug Fritz, Thibault Sottiaux, Mantas Pajarskas, Tobias Pohlen, Zhitao Gong, Daniel Toyama, Cyprien de Masson d'Autume, Yujia Li, Tayfun Terzi, Vladimir Mikulik, Igor Babuschkin, Aidan Clark, Diego de Las Casas, Aurelia Guy, Chris Jones, James Bradbury, Matthew G. Johnson, Blake A. Hechtman, Laura Weidinger, Iason Gabriel, William S. Isaac, Edward Lockhart, Simon Osindero, Laura Rimell, Chris Dyer, Oriol Vinyals, Kareem W. Ayoub, Jeff Stanway, L. L. Bennett, Demis Hassabis, Koray Kavukcuoglu, and Geoffrey Irving. 2021. Scaling language models: Methods, analysis & insights from training gopher. *ArXiv*, abs/2112.11446.

Abhilasha Ravichander, Aakanksha Naik, Carolyn Rose, and Eduard Hovy. 2019. EQUATE: A benchmark evaluation framework for quantitative reasoning in natural language inference. In *Proceedings of the 23rd Conference on Computational Natural Language Learning (CoNLL)*, pages 349–361, Hong Kong, China. Association for Computational Linguistics.

Subhro Roy, Tim Vieira, and Dan Roth. 2015. Reasoning about quantities in natural language. *Transactions of the Association for Computational Linguistics*, 3:1–13.

Georgios Spithourakis and Sebastian Riedel. 2018. Numeracy for language models: Evaluating and improving their ability to predict numbers. In *Proceedings of the 56th Annual Meeting of the Association for Computational Linguistics (Volume 1: Long Papers)*, pages 2104–2115, Melbourne, Australia. Association for Computational Linguistics.

Daniel Spokoyny and Taylor Berg-Kirkpatrick. 2020. An empirical investigation of contextualized number prediction. In *EMNLP*.

Daniel Spokoyny, Ivan Lee, Zhao Jin, and Taylor Berg-Kirkpatrick. 2022. Masked measurement prediction: Learning to jointly predict quantities and units from textual context. In *Findings of the Association for Computational Linguistics: NAACL 2022*, pages 17–29, Seattle, United States. Association for Computational Linguistics.

Aarohi Srivastava and et al. 2022. Beyond the imitation game: Quantifying and extrapolating the capabilities of language models. *ArXiv*, abs/2206.04615.

Dhanasekar Sundararaman, Vivek Subramanian, Guoyin Wang, Liyan Xu, and Lawrence Carin. 2022. Improving downstream task performance by treating numbers as entities.

Avijit Thawani, Jay Pujara, and Filip Ilievski. 2021. Numeracy enhances the literacy of language models. In *Proceedings of the 2021 Conference on Empirical Methods in Natural Language Processing*, pages 6960–6967, Online and Punta Cana, Dominican Republic. Association for Computational Linguistics.

Yan Wang, Xiaojiang Liu, and Shuming Shi. 2017. Deep neural solver for math word problems. In *Proceedings of the 2017 Conference on Empirical Methods in Natural Language Processing*, pages 845–854, Copenhagen, Denmark. Association for Computational Linguistics.

A Appendix

Task instructions

You will be shown a sentence with two numbers marked with stars (**) inside the text. Please choose the relationship between these two numbers from one of the 3 categories mentioned below.

Label categories

Positive:

If you were to increase one number you would expect the other number to also increase.
If you were to decrease one number you would expect the other number to also decrease.
Examples:

> Sentence:: My **40** liter luggage weights **50** pounds when full.
> Answer:: Answer: Positive relationship
> Explanation:: The first number describes the volume in liters of the luggage. If the volume increases we expect the weight of to also increase when it is filled up.

Negative:

If you were to increase one number you would expect the other number to decrease.
If you were to decrease one number you would expect the other number increase.
Examples:

> Sentence::He smokes **3** packs a day and his expected life age is **73**
> Answer:: Negative relationship
> Explanation:: Smoking cigarettes lowers your expected life age. Increasing the number of cigarettes you smoke should result in

No Relationship:

Increasing or decreasing one number should result in no predictable or senseable changes to the second number.
Examples:

> Sentence:: There are **200** coffee shops in Amsterdam and the average person bikes **15** miles a day.
> Answer:: No relationship
> Explanation:: Having more or fewer coffee shops may change the average amount people in Amsterdam bike but not in any readily predictable and senseable way.
>
> Sentence:: Comprising **219** sqkm of land, the city proper has **4,457** inhabitants per km2.
> Answer:: No relationship
> Explanation:: If the city has less land it may have a higher density of people, however, it may also be a smaller city that has less land, smaller population and thus less people.

Figure 5: Instructions given to the labellers for the qualification task.

# Ex	Text	Label
1.	I wear my nike shoes out in only **3** months because the soles are only **1/2** an inch thick.	Positive
2.	To cook a 20 lb turkey place in the oven for **2** hours at **435** degrees.	Negative
3.	Jordan trained for his race by running **5** miles at a pace of **10** mph.	Negative
4.	The president travels on average **thirty** times a year on Air Force one a Boeing **747**.	No Relationship
5.	My house has **2** bedrooms and is **1800** square feet.	Positive
6.	Blackthorn was one of **39** original **180** feet seagoing buoy tenders built between 1942-1944.	No Relationship
7.	The family bought a **two** ton pickup truck with 180 hp and a fuel efficiency of **25** miles per gallon.	Negative
8.	My subaru has a **4** cylinder and **150** horse power enginer.	Positive
9.	Like all Type UB III submarines UB-102 carried **10** torpedoes and was armed with a **10** cms deck gun.	No Relationship
10.	The Triple Crown of Canoe Racing consists of three separate marathon races with a total distance of **308** miles over **5** days of racing.	Positive

Table 2: The ten examples used to qualify AMTworkers.

Extracting Operator Trees from Model Embeddings

Anja Reusch and **Wolfgang Lehner**
Database Systems Group,
Technische Universität Dresden, Germany
firstname.lastname@tu-dresden.de

Abstract

Transformer-based language models are able to capture several linguistic properties such as hierarchical structures like dependency or constituency trees. Whether similar structures for mathematics are extractable from language models has not yet been explored. This work aims to probe current state-of-the-art models for the extractability of Operator Trees from their contextualized embeddings using the structure probe designed by (Hewitt and Manning, 2019). We release the code and our data set for future analyses[1].

1 Introduction

Transformer-based Language Models have not only a high impact on all domains in Natural Language Understanding but also on related fields that besides natural language try to model artificial languages such as programming code or mathematical notation written in LaTeX (Feng et al., 2020; Peng et al., 2021). The knowledge or linguistic properties that models like BERT or RoBERTa capture have been the subject of several studies: According to recent research, BERT encodes information about part-of-speech tags, roles, and syntactic features such as constituency and dependency trees (Rogers et al., 2020). Since transformer-encoder-based models were applied successfully for mathematical question answering or notation prediction (Reusch et al., 2022b; Jo et al., 2021), these models must have also acquired mathematical knowledge. However, the field of interpretability for mathematical information has not been a topic of research so far. Therefore, this work aims to analyze the prevalence of one type of mathematical knowledge: Operator Trees, a type of parse trees that can be generated from LaTeX formulas.

Generally, whether a model encodes a certain property is evaluated by applying a probe, i.e., a classifier that is trained on top of the contextualized embeddings. The performance of this classifier is used as an indicator whether the information about the property was encoded in the contextualized embeddings. To analyze whether it is possible to reconstruct an Operator Tree from the contextualized embeddings of a transformer-encoder model, we apply the structural probe introduced by (Hewitt and Manning, 2019). This probe approximates the distance between nodes in the trees using the distance of two embeddings.

In total, we train the structural probe on the embeddings of each layer of nine models for math and science and show that in most cases it is possible to reconstruct Operator Trees from the models' contextualized embeddings. The highest correlation between the learned tree distance and the gold standard is reached in the middle layers, e.g., around layer 6 for models based on bert-base and roberta-base. As Hewitt et al. also found for dependency trees, most models follow a similar pattern of information spreading among layers.

2 Related Work

Within the last years, several transformer-encoder-based models for mathematics have been developed with different applications in mind. The recent ARQMath Lab 3 (Mansouri et al., 2022) included several teams that applied models pre-trained on math: MIRMU used mathBERTa, a model based on roberta-base (Novotný and Štefánik, 2022; Geletka et al., 2022), (Reusch et al., 2022a) adapted albert-base-v2 and roberta-base for math, and (Zhong et al., 2022) further pre-trained a BERT model. In ARQMath Lab 1, the team PSU also released a further pre-trained model based on RoBERTa (Rohatgi et al., 2020). In addition, (Jo et al., 2021) fine-tuned a BERT model for notation prediction tasks based on scientific documents. MathBERT (Peng et al., 2021) leverages operator trees during pre-training for several tasks such as formula topic

[1] https://github.com/AnReu/extracting-opts

Proceedings of the 1st Workshop on Mathematical Natural Language Processing (MathNLP), pages 40 - 50
December 8, 2022 ©2022 Association for Computational Linguistics

classification and information retrieval. Related to mathematics is also the domain of scientific documents for which SciBERT was trained (Beltagy et al., 2019).

However, little is known so far about what BERT-based models learn about mathematics. In contrast, their learning capacities on natural language received large attention in recent research (for a survey see (Rogers et al., 2020)). Several probes and classifiers were employed to analyze whether BERT captures grammatical structures like dependency or constituency trees (Tenney et al., 2019; Hewitt and Manning, 2019; Coenen et al., 2019) or which layer attends to which linguistic feature (Clark et al., 2019). Visual frameworks like bertviz by (Vig, 2019) support the analysis of BERT's inner working by visualizing the attention weights of trained models. Also ALBERT was shown to capture part-of-speech tags in different places as reported by (Chiang et al., 2020), but most studies were performed using BERT.

3 Probing for Mathematical Structures

We analyze whether it is possible to reconstruct mathematical parse trees from the models' contextualized embeddings. It was already shown that BERT is able to learn grammatical structures of natural languages which could be extracted in the form of constituency and dependency trees (Tenney et al., 2019; Hewitt and Manning, 2019). Therefore, we apply the same type of probe to test for Operator Trees.

3.1 Structural Probe

The goal of the structural probe as introduced by (Hewitt and Manning, 2019) is to learn a matrix B, such that the distance d_B defined by $d_B(U_i, U_j) := \sqrt{(U_i - U_j)^\mathsf{T} B^\mathsf{T} B (U_i - U_j)}$ approximates a given tree distance d_T, i.e., the length of the path between the node of word s_i and the one of word s_j in the tree of example s. U_i and U_j are the contextualized embeddings of the words s_i and s_j in s. B is learned by minimizing the loss function over each examples $s \in S$ in the training corpus:

$$\min_B \sum_{s \in S} \frac{1}{|s|^2} \sum_{i,j} |d_T(s_i, s_j) - d_B(U_i, U_j)|$$

Originally, Hewitt et al. applied the structural probe to demonstrate that dependency structures of the English language are, to some extent, contained in BERT's contextualized embeddings. In this work, we will train a structure probe to evaluate whether the models' inner workings have learned about mathematical structures, i.e., operator trees.

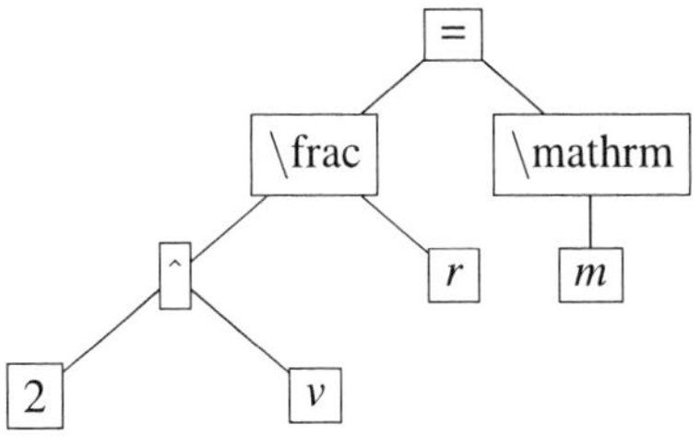

Figure 1: Operator Tree of the formula $\mathrm{m} = \frac{r}{v^2}$

3.2 Operator Trees

Formulas possess a hierarchical structure, which is encoded in Content Math ML [2], defining an operator tree (OPT). An example OPT for the equation $\mathrm{m} = \frac{r}{v^2}$ is shown in Fig. 1. Nodes of this tree representation can be individual or multiple symbols such as numbers, variables, text fragments indicating certain functions, fractions, radicals, LaTeX style expressions, or parentheses and brackets. This definition is similar to the one found in (Mansouri et al., 2019), but we added parentheses and brackets to investigate the way the models capture open and closed bracket relationships. OPT edges indicate an operator-argument relationship between parent and child nodes. Left and right brackets and parentheses have each an edge to the parent of the tree inside them. LaTeX style expressions like \mathbb can be simply seen as an operator applied on the argument inside. Hence, the original OPT stays intact.

4 Experimental Setup

We evaluated in total 13 models which are publicly available on the Huggingface Model Hub[3]. We chose the eight mathematical models by searching the Model Hub for transformer-encoder models that were (further) pre-trained on mathematics. We also added the popular model SciBERT as its domain, science, is close to mathematics. In addition, the four models which served as a base for pre-training were evaluated. A summary of the models can be found in Tab. 1. Of particular interest would have been an evaluation of MathBERT by (Peng et al., 2021), a model that relied on Operator Trees during

[2] https://w3c.github.io/mathml/#contm
[3] https://huggingface.co/models

Model Identifier	Base Model	Data Set
albert-base-v2	-	Books and Wikipedia
AnReu/math_albert	albert-base-v2	ARQMath
bert-base-cased	-	Books and Wikipedia
allenai/scibert-scivocab-cased	-	Scientific documents
AnReu/math_pretrained_bert	bert-base-cased	ARQMath
tbs17/MathBERT	-	Math text books, curricula, paper abstracts
tbs17/ MathBERT-custom	-	Math text books, curricula, paper abstracts
roberta-base	-	Books, Wikipedia, news, websites, stories
roberta-large	-	Books, Wikipedia, news, websites, stories
AnReu/math_pretrained_roberta	roberta-base	ARQMath
shauryr/arqmath-roberta-base	roberta-base	ARQMath
uf-aice-lab/math-roberta	roberta-large	Math discussion posts
witiko/mathberta	roberta-base	ARQMath, ArXiv documents

Table 1: Summary of the evaluated models, their base models and the data sets used for pre-training.

pre-training. However, neither the model nor the code are publicly available.

BERT, ALBERT and RoBERTa were trained on a general natural language corpus and serve as baselines. Six models were further pre-trained from a base model like BERT or RoBERTa, while three were developed from scratch. The data sources the models were trained on are rather diverse: Five models use ARQMath, others use math text books, school curricula, paper abstracts, or other discussion posts apart from ARQMath. SciBERT is the only model what was not specifically trained on mathematical content, but on scientific publications. All models can be found on Huggingface by using their model identifier.

4.1 Data

Our probe is trained on formulas, which were parsed to OPTs by a custom LaTeX parser written in Python adapted from the parser rules of the mathematical formula search engine Approach0 (Zhong and Zanibbi, 2019; Zhong et al., 2020). We could not use existing parsers because it is necessary to associate each LaTeX token with its node in the OPT and existing parsers only output the entire parse tree without annotation of a node's token in the formula. We selected 50k training examples by chance from the corpus of all formulas from ARQMath 2020 (Mansouri et al., 2020), which contains question and answer posts from the Q&A community Mathematics StackExchange[4]. From the remaining set we chose 10k for development, and an additional set of 10k as test set. The average number of nodes in all three sets is 16.5, while the average tree depth is 4.8. The most common node types are variables and numbers, followed by LaTeX braces and relation symbols. Among the relation symbols, the equal sign "=" occurs most often.

4.2 Metrics

We follow Hewitt et al. and evaluate the performance using UUAS (Undirected Unlabeled Attachment Score), which denotes the percentage of correctly identified edges in the predicted tree and, distance Spearman (DSpr.), which is determined by first calculating the Spearman correlation between the predicted distances d_B and the gold-standard distances d_T. These correlations are then averaged among all formulas of a fixed length. Finally, the average across formulas of lengths 5–50 is reported as DSpr. We decided to include both metrics since it was shown that their scores can result in opposite trends (Hall Maudslay et al., 2020).

4.3 Setup

To train and evaluate the probing classifier, we used the original code provided by Hewitt et al.[5] and adapted it to the transformers library[6]. We used the L1 loss and a maximum rank of the probe of 768, as reported by the authors. We trained the probes

[4]https://math.stackexchange.com

[5]https://github.com/john-hewitt/structural-probes

[6]https://pypi.org/project/transformers/

using one A100 GPU with 40 GB GPU memory. Depending on the base model, the training of a probe took between 15 min and 1.5h. Each model was trained on five different random seeds.

5 Results

Tab. 2 summarizes the highest values from all layers. We report our results using UUAS and DSpr. where higher values indicate a larger percentage of correctly reconstructed edges and a higher correlation between the predicted and gold-distances, respectively. Each value is the mean of the five runs. It is visible that almost all adapted models improve over their natural language baselines. The highest performance overall is demonstrated by AnReu/math_pretrained_bert. Only the performance of MathBERT-custom drops in comparison to bert-base-cased. The DSpr. scores of the best models in comparison to their baselines are visualized in Fig. 2.

In general, the models pre-trained on ARQMath demonstrated a better performance across both metrics compared to models pre-trained on other data sets. A possible reason could be that this data set contains a large variety of formulas written in LaTeX while this is unclear for the other data sets since they are not publicly available. We validated these results also using a second OPT data set based on the MATH data set (Hendrycks et al., 2021), which contains formulas written in LaTeXextracted from competition math problems. Since there was no drop in performance among the models pre-trained on ARQMath, we can conclude that models did not benefit from the overlap between the pre-training data and the probing formulas.

BERT and RoBERTa-based models show that the best extractability for Operator Trees lies in the middle layers, between layer 4 and 7 for base models and between layer 9 and 13 for large models. This pattern is consistent with the results reported by Hewitt et al. for dependency structures. Notably, the same pattern does not emerge for ALBERT and AnReu/math_albert. Here, the highest scores are in layers 2 and 3. Overall, the scores for both ALBERT-based models are significantly lower, even after training on ARQMath. Interestingly, this model was among the best for the ARQMath Lab 3 on Mathematical Answer Retrieval and outperformed also AnReu/math_pretrained_roberta, which is the second best model for UUAS in this study. A similar mismatch between the performance in downstream natural language tasks and syntactic parsing was also found by (Glavaš and Vulić, 2021). Therefore, this finding casts a doubt on whether the models rely on their OPT knowledge when solving the downstream task of Mathematical Answer Retrieval. However, the limitations of probing classifiers as the one used in this work do not allow conclusions about the models usage of the knowledge. Hence, further research in this direction is required to investigate whether and how these models use structural knowledge during downstream tasks. In addition, Appendix A shows examples of reconstructed Operator Trees, while Appendix B contains the mean scores and standard deviation for each model in each layer.

Model	DSpr.	UUAS
albert-base-v2	0.631 (3)	0.477 (3)
AnReu/math_albert	0.680 (2)	0.513 (2)
bert-base-cased	0.713 (7)	0.532 (6)
allenai/ scibert-scivocab-cased	0.727 (7)	0.545 (7)
AnReu/ math_pretrained_bert	**0.815** (7)	**0.700** (6)
tbs17/MathBERT	0.718 (6)	0.550 (5)
tbs17/ MathBERT-custom	0.686 (5)	0.530 (5)
roberta-base	0.703 (5)	0.526 (5)
roberta-large	0.706 (9)	0.536 (13)
AnReu/ math_pretrained_roberta	0.746 (5)	0.576 (5)
shauryr/ arqmath-roberta-base	0.715 (5)	0.541 (4)
uf-aice-lab/ math-roberta	0.711 (9)	0.547 (11)
witiko/mathberta	0.752 (5)	0.574 (5)

Table 2: Results of reconstruction of OPTs using UUAS and DSpr., displaying only the best results across all layers, best layer indicated by (layer number).

6 Conclusion

This work aims to answer the question: Are Operator Trees extractable from the models' contextualized embeddings? We trained a structural probe that learns to approximate the distances between nodes in the trees. The results show that models (further) pre-trained on mathematical data sets outperform their natural language baselines. The high correlation of the trained probe suggests that the

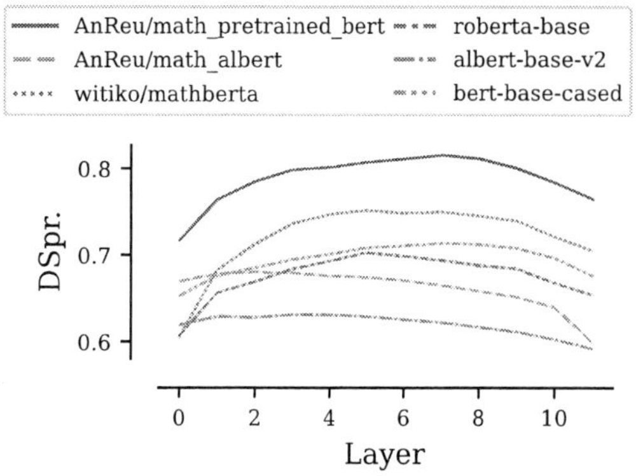

Figure 2: Results of reconstruction of OPTs across layers of the best models and all baselines, results for DSpr., the same pattern emerges for UUAS.

models indeed encode useful information about Operator Trees in their contextualized embeddings. Given that the models have never been trained on Operator Trees, but only using masked-language modeling on string-based representation such as LaTeX, explicitly proving Operator Trees during a downstream task such as Mathematical Retrieval might not even be necessary.

Furthermore, we notice differences between model classes: While BERT and RoBERTa-based models demonstrate a higher extractability for the structural probe, both ALBERT-based models fall behind. In contrast, their performance on mathematical answer retrieval is on par with the other evaluated models. Further research is required to investigate this issue. We are open to offer other researchers the re-use of our work by making our source code and data set fully publicly available on GitHub[7].

Acknowledgements

The authors would like to thank the anonymous reviewers for their helpful feedback and comments. This work was supported by the DFG under Germany's Excellence Strategy, Grant No. EXC-2068-390729961, Cluster of Excellence "Physics of Life" of TU Dresden. Furthermore, the authors are grateful for the GWK support for funding this project by providing computing time through the Center for Information Services and HPC (ZIH) at TU Dresden.

[7]`https://github.com/AnReu/extracting-opts`

References

Iz Beltagy, Kyle Lo, and Arman Cohan. 2019. Scibert: A pretrained language model for scientific text. In *Proceedings of the 2019 Conference on Empirical Methods in Natural Language Processing and the 9th International Joint Conference on Natural Language Processing (EMNLP-IJCNLP)*, pages 3606–3611.

Cheng-Han Chiang, Sung-Feng Huang, and Hung-yi Lee. 2020. Pretrained language model embryology: The birth of albert. In *Proceedings of the 2020 Conference on Empirical Methods in Natural Language Processing (EMNLP)*, pages 6813–6828.

Kevin Clark, Urvashi Khandelwal, Omer Levy, and Christopher D Manning. 2019. What does bert look at? an analysis of bert's attention. *arXiv preprint arXiv:1906.04341*.

Andy Coenen, Emily Reif, Ann Yuan, Been Kim, Adam Pearce, Fernanda Viégas, and Martin Wattenberg. 2019. Visualizing and measuring the geometry of bert. In *Proceedings of the 33rd International Conference on Neural Information Processing Systems*, pages 8594–8603.

Zhangyin Feng, Daya Guo, Duyu Tang, Nan Duan, Xiaocheng Feng, Ming Gong, Linjun Shou, Bing Qin, Ting Liu, Daxin Jiang, et al. 2020. Codebert: A pre-trained model for programming and natural languages. *arXiv preprint arXiv:2002.08155*.

Martin Geletka, Vojtěch Kalivoda, Michal Štefánik, Marek Toma, and Petr Sojka. 2022. Diverse semantics representation is king.

Goran Glavaš and Ivan Vulić. 2021. Is supervised syntactic parsing beneficial for language understanding tasks? an empirical investigation. In *Proceedings of the 16th Conference of the European Chapter of the Association for Computational Linguistics: Main Volume*, pages 3090–3104.

Rowan Hall Maudslay, Josef Valvoda, Tiago Pimentel, Adina Williams, and Ryan Cotterell. 2020. A tale of a probe and a parser. In *Proceedings of the 58th Annual Meeting of the Association for Computational Linguistics*, pages 7389–7395, Online. Association for Computational Linguistics.

Dan Hendrycks, Collin Burns, Saurav Kadavath, Akul Arora, Steven Basart, Eric Tang, Dawn Song, and Jacob Steinhardt. 2021. Measuring mathematical problem solving with the math dataset. *NeurIPS*.

John Hewitt and Christopher D Manning. 2019. A structural probe for finding syntax in word representations. In *Proceedings of the 2019 Conference of the North American Chapter of the Association for Computational Linguistics: Human Language Technologies, Volume 1 (Long and Short Papers)*, pages 4129–4138.

Hwiyeol Jo, Dongyeop Kang, Andrew Head, and Marti A Hearst. 2021. Modeling mathematical notation semantics in academic papers. In *Findings of the*

Association for Computational Linguistics: EMNLP 2021, pages 3102–3115.

Behrooz Mansouri, Anurag Agarwal, Douglas Oard, and Richard Zanibbi. 2020. Finding old answers to new math questions: the arqmath lab at clef 2020. In *European Conference on Information Retrieval*, pages 564–571. Springer.

Behrooz Mansouri, Vít Novotný, Anurag Agarwal, Douglas W Oard, and Richard Zanibbi. 2022. Overview of arqmath-3 (2022): Third clef lab on answer retrieval for questions on math (working notes version). *Working Notes of CLEF*.

Behrooz Mansouri, Shaurya Rohatgi, Douglas W Oard, Jian Wu, C Lee Giles, and Richard Zanibbi. 2019. Tangent-cft: An embedding model for mathematical formulas. In *Proceedings of the 2019 ACM SIGIR international conference on theory of information retrieval*, pages 11–18.

Vít Novotný and Michal Štefánik. 2022. Combining sparse and dense information retrieval. *Proceedings of the Working Notes of CLEF*.

Shuai Peng, Ke Yuan, Liangcai Gao, and Zhi Tang. 2021. Mathbert: A pre-trained model for mathematical formula understanding. *arXiv preprint arXiv:2105.00377*.

Anja Reusch, Maik Thiele, and Wolfgang Lehner. 2022a. Transformer-encoder and decoder models for questions on math. *Proceedings of the Working Notes of CLEF 2022*, pages 5–8.

Anja Reusch, Maik Thiele, and Wolfgang Lehner. 2022b. Transformer-encoder-based mathematical information retrieval. In *International Conference of the Cross-Language Evaluation Forum for European Languages*, pages 175–189. Springer.

Anna Rogers, Olga Kovaleva, and Anna Rumshisky. 2020. A primer in bertology: What we know about how bert works. *Transactions of the Association for Computational Linguistics*, 8:842–866.

Shaurya Rohatgi, Jian Wu, and C Lee Giles. 2020. Psu at clef-2020 arqmath track: Unsupervised re-ranking using pretraining. In *CLEF (Working Notes)*.

Ian Tenney, Patrick Xia, Berlin Chen, Alex Wang, Adam Poliak, R Thomas McCoy, Najoung Kim, Benjamin Van Durme, Samuel R Bowman, Dipanjan Das, et al. 2019. What do you learn from context? probing for sentence structure in contextualized word representations. *arXiv preprint arXiv:1905.06316*.

Jesse Vig. 2019. A multiscale visualization of attention in the transformer model. *arXiv preprint arXiv:1906.05714*.

Wei Zhong, Shaurya Rohatgi, Jian Wu, C Lee Giles, and Richard Zanibbi. 2020. Accelerating substructure similarity search for formula retrieval. *Advances in Information Retrieval*, 12035:714.

Wei Zhong, Yuqing Xie, and Jimmy Lin. 2022. Applying structural and dense semantic matching for the arqmath lab 2022, clef. *Proceedings of the Working Notes of CLEF 2022*, pages 5–8.

Wei Zhong and Richard Zanibbi. 2019. Structural similarity search for formulas using leaf-root paths in operator subtrees. In *European Conference on Information Retrieval*, pages 116–129. Springer.

A Examples of Reconstructed Operator Trees

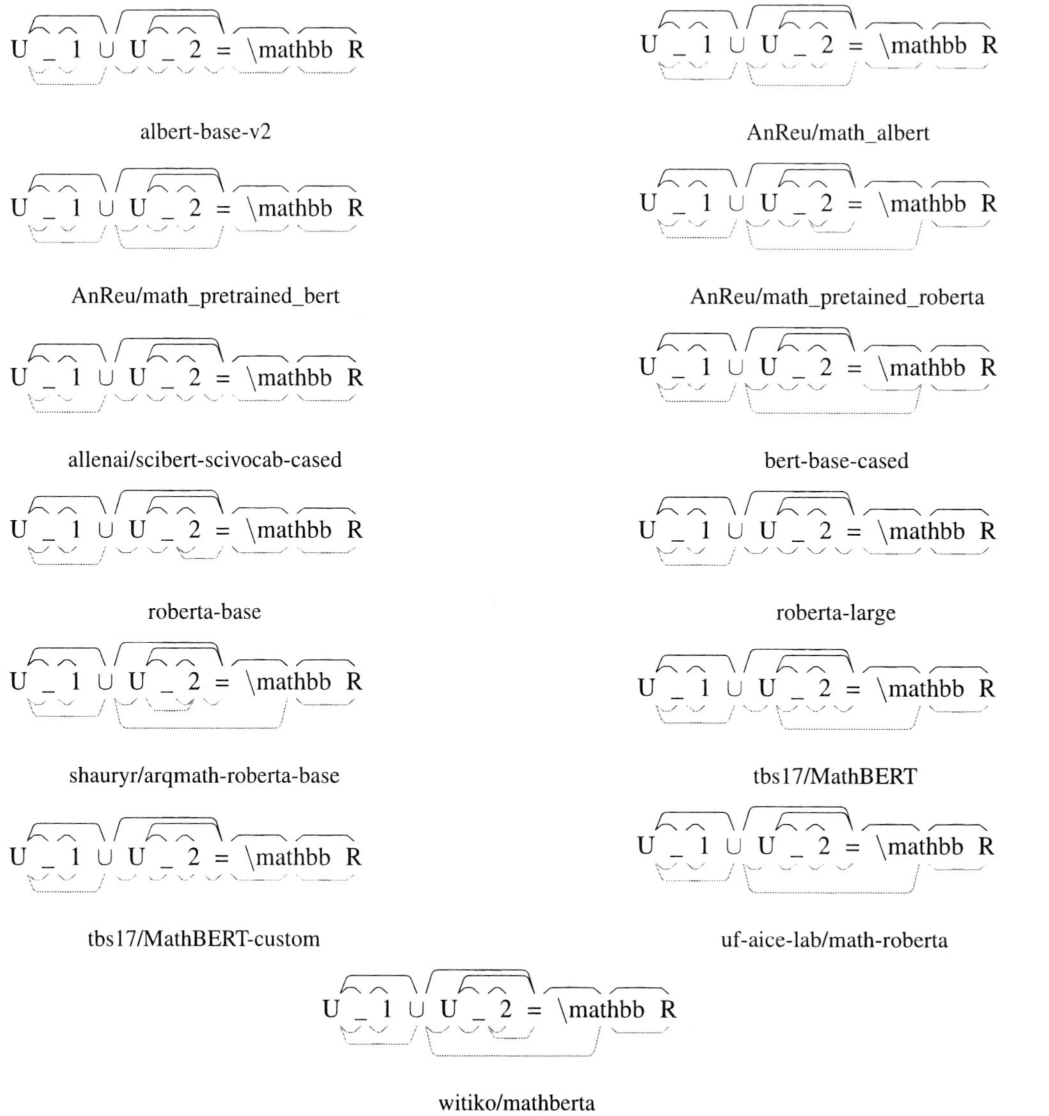

Figure 3: Operator Trees calculated from the predicted squared distances between the tokens. The black edges above each formula are the gold edges from the OPT parser, while the red edges are the predicted ones by each model, taken from one seed of the best layer by DSpr. In a large majority of cases the models correctly identified the edges of the displayed formula. Most differences can be seen from the second part of the left hand side of the equation, where the models mostly struggle with the parent-child relationships of the equal sign.

B Results for all layers

	bert-base-cased		tbs17/MathBERT		tbs17/MathBERT-custom	
	mean	stdev	mean	stdev	mean	stdev
0	0.6525	0.00126	0.6455	0.00040	0.6468	0.00058
1	0.6759	0.00116	0.6481	0.00110	0.6492	0.00035
2	0.6851	0.00075	0.6540	0.00084	0.6530	0.00078
3	0.6949	0.00078	0.6834	0.00047	0.6785	0.00080
4	0.7008	0.00045	0.7047	0.00083	0.6824	0.00049
5	0.7082	0.00027	0.7145	0.00036	0.6863	0.00073
6	0.7106	0.00019	0.7175	0.00036	0.6832	0.00009
7	0.7134	0.00037	0.7092	0.00044	0.6772	0.00024
8	0.7121	0.00042	0.6987	0.00026	0.6703	0.00016
9	0.7079	0.00033	0.6842	0.00028	0.6609	0.00027
10	0.6965	0.00013	0.6646	0.00021	0.6531	0.00031
11	0.6759	0.00046	0.6495	0.00026	0.6425	0.00031
	allenai/scibert_scivocab_cased		AnReu/math_pretrained_bert			
0	0.6578	0.00094	0.7167	0.00020		
1	0.6835	0.00122	0.7639	0.00032		
2	0.6962	0.00110	0.7848	0.00041		
3	0.7061	0.00033	0.7985	0.00030		
4	0.7200	0.00040	0.8015	0.00011		
5	0.7263	0.00046	0.8070	0.00017		
6	0.7267	0.00009	0.8110	0.00005		
7	0.7261	0.00016	**0.8154**	0.00012		
8	0.7150	0.00063	0.8116	0.00006		
9	0.6938	0.00008	0.8007	0.00018		
10	0.6792	0.00019	0.7839	0.00008		
11	0.6764	0.00031	0.7647	0.00010		

Table 3: DSpr. Results of BERT, BERT-based and similar models.

	bert-base-cased		tbs17/MathBERT		tbs17/MathBERT-custom	
	mean	stdev	mean	stdev	mean	stdev
0	0.4585	0.00169	0.4832	0.00113	0.4891	0.00053
1	0.4947	0.00107	0.4845	0.00129	0.4927	0.00061
2	0.5109	0.00068	0.4845	0.00059	0.4929	0.00043
3	0.5178	0.00027	0.5171	0.00028	0.5162	0.00049
4	0.5216	0.00059	0.5393	0.00051	0.5255	0.00074
5	0.5315	0.00064	0.5496	0.00037	0.5300	0.00059
6	0.5323	0.00019	0.5491	0.00039	0.5248	0.00044
7	0.5321	0.00053	0.5363	0.00038	0.5169	0.00050
8	0.5283	0.00040	0.5202	0.00043	0.5074	0.00042
9	0.5221	0.00051	0.5017	0.00018	0.4973	0.00045
10	0.5032	0.00018	0.4756	0.00022	0.4854	0.00038
11	0.4779	0.00035	0.4560	0.00028	0.4725	0.00034
	allenai/scibert_scivocab_cased		AnReu/math_pretrained_bert			
0	0.4655	0.00036	0.5458	0.00069		
1	0.4984	0.00150	0.6336	0.00054		
2	0.5151	0.00103	0.6686	0.00034		
3	0.5244	0.00037	0.6825	0.00030		
4	0.5363	0.00038	0.6790	0.00043		
5	0.5450	0.00038	0.6920	0.00032		
6	0.5421	0.00066	**0.7000**	0.00043		
7	0.5453	0.00018	0.6952	0.00041		
8	0.5308	0.00087	0.6852	0.00043		
9	0.5101	0.00036	0.6694	0.00046		
10	0.4925	0.00033	0.6415	0.00030		
11	0.4865	0.00038	0.6028	0.00046		

Table 4: UUAS Results of BERT, BERT-based and similar models.

	albert-base-v2		AnReu/math_albert	
	mean	stdev	mean	stdev
0	0.6192	0.00072	0.6693	0.00017
1	0.6290	0.00133	0.6783	0.00022
2	0.6279	0.00039	**0.6805**	0.00043
3	0.6312	0.00030	0.6791	0.00024
4	0.6310	0.00049	0.6759	0.00015
5	0.6291	0.00029	0.6743	0.00031
6	0.6255	0.00067	0.6706	0.00014
7	0.6221	0.00082	0.6649	0.00017
8	0.6168	0.00053	0.6583	0.00017
9	0.6115	0.00041	0.6516	0.00041
10	0.6028	0.00037	0.6397	0.00033
11	0.5919	0.00040	0.5991	0.00025

(a) DSpr. Results

	albert-base-v2		AnReu/math_albert	
	mean	stdev	mean	stdev
0	0.4620	0.00128	0.5095	0.00039
1	0.4727	0.00132	**0.5130**	0.00040
2	0.4746	0.00054	0.5125	0.00025
3	0.4771	0.00055	0.5127	0.00038
4	0.4742	0.00063	0.5090	0.00032
5	0.4747	0.00059	0.5062	0.00023
6	0.4699	0.00088	0.5030	0.00062
7	0.4652	0.00091	0.4975	0.00048
8	0.4563	0.00070	0.4891	0.00032
9	0.4470	0.00049	0.4811	0.00024
10	0.4343	0.00088	0.4671	0.00056
11	0.4144	0.00075	0.4082	0.00046

(b) UUAS Results

Figure 4: Results of ALBERT and math albert.

	roberta-base		shauryr/ arqmath-roberta-base		witiko/mathberta		AnReu/ math_pretrained_roberta	
	mean	stdev	mean	stdev	mean	stdev	mean	stdev
0	0.6066	0.00085	0.6219	0.00054	0.6051	0.00024	0.6179	0.00036
1	0.6561	0.00022	0.6682	0.00048	0.6824	0.00065	0.6917	0.00038
2	0.6691	0.00039	0.6841	0.00049	0.7123	0.00117	0.7195	0.00057
3	0.6840	0.00044	0.7014	0.00085	0.7363	0.00043	0.7345	0.00038
4	0.6930	0.00027	0.7114	0.00045	0.7475	0.00021	0.7422	0.00027
5	0.7025	0.00026	0.7146	0.00027	**0.7519**	0.00065	0.7464	0.00030
6	0.6986	0.00053	0.7102	0.00019	0.7487	0.00069	0.7409	0.00005
7	0.6937	0.00067	0.7038	0.00015	0.7501	0.00022	0.7366	0.00041
8	0.6881	0.00093	0.6997	0.00027	0.7456	0.00035	0.7331	0.00020
9	0.6843	0.00058	0.6961	0.00018	0.7399	0.00015	0.7274	0.00027
10	0.6677	0.00061	0.6798	0.00050	0.7207	0.00020	0.7094	0.00026
11	0.6538	0.00036	0.6616	0.00031	0.7049	0.00034	0.6942	0.00024

Table 5: DSpr. Results of RoBERTa-base and small RoBERTA-based models.

	roberta-base		shauryr/ arqmath-roberta-base		witiko/mathberta		AnReu/ math_pretrained_roberta	
	mean	stdev	mean	stdev	mean	stdev	mean	stdev
0	0.4456	0.00127	0.4619	0.00044	0.4268	0.00078	0.4543	0.00068
1	0.4825	0.00042	0.5010	0.00083	0.5086	0.00068	0.5296	0.00084
2	0.4988	0.00053	0.5184	0.00053	0.5388	0.00065	0.5477	0.00095
3	0.5187	0.00015	0.5352	0.00059	0.5618	0.00065	0.5690	0.00063
4	0.5224	0.00042	0.5414	0.00061	0.5732	0.00027	0.5731	0.00043
5	0.5255	0.00041	0.5378	0.00017	**0.5742**	0.00084	0.5759	0.00026
6	0.5172	0.00053	0.5358	0.00026	0.5687	0.00057	0.5672	0.00024
7	0.5088	0.00035	0.5247	0.00059	0.5692	0.00038	0.5591	0.00029
8	0.5106	0.00033	0.5311	0.00061	0.5704	0.00055	0.5599	0.00026
9	0.5091	0.00057	0.5324	0.00016	0.5659	0.00025	0.5590	0.00041
10	0.4899	0.00032	0.5167	0.00033	0.5471	0.00043	0.5411	0.00027
11	0.4713	0.00051	0.4902	0.00030	0.5294	0.00035	0.5238	0.00037

Table 6: UUAS Results of RoBERTa-base and small RoBERTa-based models.

	roberta-large		uf-aice-lab/ math-roberta	
	mean	stdev	mean	stdev
0	0.61374	0.00029	0.61387	0.00051
1	0.62816	0.00057	0.62524	0.00068
2	0.65153	0.00077	0.65213	0.00080
3	0.66078	0.00116	0.66054	0.00058
4	0.67341	0.00050	0.67043	0.00083
5	0.68048	0.00111	0.68125	0.00034
6	0.68313	0.00044	0.68719	0.00052
7	0.68996	0.00057	0.69613	0.00054
8	0.69757	0.00072	0.70321	0.00068
9	0.70602	0.00058	**0.71079**	0.00046
10	0.70484	0.00029	0.70922	0.00043
11	0.70425	0.00061	0.70943	0.00068
12	0.70255	0.00106	0.70796	0.00041
13	0.70144	0.00057	0.70940	0.00042
14	0.69807	0.00035	0.70646	0.00044
15	0.69522	0.00046	0.70268	0.00048
16	0.69463	0.00012	0.70220	0.00032
17	0.69444	0.00023	0.70017	0.00025
18	0.68966	0.00009	0.69823	0.00025
19	0.68492	0.00029	0.69437	0.00014
20	0.68087	0.00036	0.69277	0.00027
21	0.67582	0.00035	0.69129	0.00012
22	0.66197	0.00051	0.68796	0.00057
23	0.64475	0.00056	0.68602	0.00082
24	0.62023	0.00042	0.66011	0.00017

(a) DSpr. Results

	roberta-large		uf-aice-lab/ math-roberta	
	mean	stdev	mean	stdev
0	0.44171	0.000279	0.44210	0.000778
1	0.45419	0.001066	0.45024	0.001192
2	0.48857	0.000532	0.48968	0.001104
3	0.50158	0.000647	0.50030	0.000336
4	0.52109	0.000171	0.51772	0.000599
5	0.51957	0.001377	0.52384	0.000602
6	0.51196	0.000350	0.52088	0.000902
7	0.52005	0.000604	0.52933	0.000500
8	0.52343	0.001005	0.53698	0.000465
9	0.53249	0.000754	0.54388	0.000493
10	0.53287	0.000664	0.54576	0.000411
11	0.53620	0.000540	**0.54715**	0.000436
12	0.53255	0.001094	0.54254	0.000349
13	0.53622	0.000277	0.54502	0.000427
14	0.52932	0.000924	0.53911	0.000366
15	0.52427	0.000548	0.53527	0.000632
16	0.52411	0.000432	0.53579	0.000367
17	0.52099	0.000341	0.53276	0.000112
18	0.51146	0.000397	0.52748	0.000328
19	0.50598	0.000293	0.52150	0.000432
20	0.50340	0.000510	0.52375	0.000435
21	0.49970	0.000665	0.52158	0.000291
22	0.48542	0.000576	0.52027	0.000370
23	0.47364	0.000890	0.52066	0.000748
24	0.44753	0.000611	0.49027	0.000578

(b) UUAS Results

Figure 5: Results of RoBERTa-large and uf-aice-lab/math-roberta.

End-to-End Evaluation of a Spoken Dialogue System for Learning Basic Mathematics

Eda Okur
Intel Labs, USA
eda.okur@intel.com

Saurav Sahay
Intel Labs, USA
saurav.sahay@intel.com

Roddy Fuentes Alba
Intel Labs, Mexico
roddy.fuentes.alba@intel.com

Lama Nachman
Intel Labs, USA
lama.nachman@intel.com

Abstract

The advances in language-based Artificial Intelligence (AI) technologies applied to build educational applications can present AI for social-good opportunities with a broader positive impact. Across many disciplines, enhancing the quality of mathematics education is crucial in building critical thinking and problem-solving skills at younger ages. Conversational AI systems have started maturing to a point where they could play a significant role in helping students learn fundamental math concepts. This work presents a task-oriented Spoken Dialogue System (SDS) built to support play-based learning of basic math concepts for early childhood education. The system has been evaluated via real-world deployments at school while the students are practicing early math concepts with multimodal interactions. We discuss our efforts to improve the SDS pipeline built for math learning, for which we explore utilizing MathBERT (Shen et al., 2021b) representations for potential enhancement to the Natural Language Understanding (NLU) module. We perform an end-to-end evaluation using real-world deployment outputs from the Automatic Speech Recognition (ASR), Intent Recognition, and Dialogue Manager (DM) components to understand how error propagation affects the overall performance in real-world scenarios.

1 Introduction

Following the advances in Artificial Intelligence (AI) research, building innovative applications to support education can present exciting opportunities with a positive and broader social impact. The United Nations (UN) Sustainable Development Goals[1] (SDGs) (Desa et al., 2016) represent an urgent call for action to help address critical global problems, where education is among the top five of these development areas (i.e., poverty, hunger, health, education, gender equality). Across many disciplines, improving the quality of mathematics education is crucial in building critical thinking and problem-solving skills at younger ages, which is a fundamental component of comprehensive and successful STEM education (i.e., science, technology, engineering, and mathematics). Language-based AI systems are starting to mature to a point where they could play a significant role in helping students learn and practice mathematical concepts. Despite its importance, applied Natural Language Processing (NLP) technologies for enhancing mathematics education still remain a highly under-explored area of research.

This study presents a task-oriented Spoken Dialogue System (SDS) developed to facilitate play-based learning of basic mathematical concepts for early childhood education. The system has been developed in the lab and evaluated via real-world deployments at school while the students are learning and practicing basic math concepts with multimodal interactions. These fundamental early math concepts and basic operations include constructing numbers using ones and tens, counting, addition, subtraction, measurement of length and size, etc. The multimodal interactions involve speech-based interactions to answer early math-related or game-related questions, counting and placing learning-specific tangible objects (i.e., manipulatives) in a visually observed space, touch-based interactions with the 1-to-100 number grid projected on the wall, to name a few.

This work discusses our efforts to improve the modular SDS pipeline built for game-based math learning and perform an end-to-end evaluation with various SDS components using real-world deployment outputs. The main SDS module we focus on investigating is Natural Language Understanding (NLU). The NLU arguably is the most critical component of goal-oriented dialogue systems that enables efficient communication between humans and intelligent conversational agents via application-

[1] https://sdgs.un.org/goals

Proceedings of the 1st Workshop on Mathematical Natural Language Processing (MathNLP), pages 51 - 64
December 8, 2022 ©2022 Association for Computational Linguistics

specific comprehensive sub-tasks. Intent Recognition (IR) is at the heart of these NLU tasks, where the goal is to identify users' objectives from the input text and determine their intentions.

The representation of human language is a crucial factor determining the success of conversational agents, especially in real-world applications. In the math learning domain, this language representation gains further significance due to the quite peculiar nature of mathematical language. With that motivation, we explored employing MathBERT (Shen et al., 2021b) representations for potential enhancement to the NLU module of our play-based math-learning system. MathBERT is a recently proposed language model created by pre-training the well-known BERT-base model (Devlin et al., 2019) on a large mathematical corpus. We compared the NLU results obtained by simply using BERT (Devlin et al., 2019) and ConveRT (Henderson et al., 2020) representations versus the new MathBERT representations. We further investigated the two variations of MathBERT models, one pre-trained with mathematics-specific vocabulary and the other with BERT-base vocabulary, to see their effects on our domain-specific math-learning NLU task.

As most of the application-specific and modular SDS pipelines do, our task-oriented SDS contains particular building blocks or modules for Automatic Speech Recognition (ASR), Natural Language Understanding (NLU), multimodal Dialogue Management (DM), Natural Language Generation (NLG), and Text-to-Speech (TTS). At its current stage of research & development, for practical reasons, we employ template-based responses at the NLG module with off-the-shelf TTS. Thus, this study emphasizes more on speech recognition (ASR), intent understanding (NLU), and response selection (DM) tasks for a conversational agent that supports elementary math learning. For a complete end-to-end evaluation of our task-oriented SDS, we evaluated the ASR, NLU, and DM components to understand how error propagation affects the overall performance in real-world scenarios.

2 Related Work

2.1 NLP for Mathematics Education

Exploring the advancements in AI systems for social good and positive impact in the education domain, specifically to amplify students' learning experiences, has recently gained increasing interest from the research community (D'Mello and Graesser, 2013; Chassignol et al., 2018; Jia et al., 2020; Baker, 2021; Zhai et al., 2021). Intelligent and interactive play-based learning systems have shown remarkable benefits for teaching mathematical concepts in smart and collaborative spaces (Lester et al., 2013; Pires et al., 2019; Richey et al., 2021; Sun et al., 2021). For early childhood education, a recent study by Skene et al. (2022) has showcased that game-based learning environments can offer significant advantages over traditional learning approaches while practicing fundamental math concepts, especially for younger learners.

Harnessing NLP technologies to construct innovative applications for education is gaining popularity as an emerging area of research having various examples in the last decade (Blanchard et al., 2015; Lende and Raghuwanshi, 2016; Taghipour and Ng, 2016; Raamadhurai et al., 2019; Cahill et al., 2020; Chan et al., 2021; Dutta et al., 2022). Within those efforts, exploring conversational agents for intelligent tutoring systems and smart education applications is a glaring sub-field of NLP in education, having several research studies tackling the problem from different angles (Graesser et al., 2004; Kerry et al., 2009; Winkler and Söllner, 2018; Palasundram et al., 2019; Winkler et al., 2020; Datta et al., 2020; Zhang et al., 2021).

Furthermore, there have been recent attempts to bridge the gap between general AI research and mathematics education (Davila and Zanibbi, 2017; Jiang et al., 2018; Mansouri et al., 2019; Yuan et al., 2020; Huang et al., 2021; Kumar and Rajagopal, 2021). To further narrow our attention to language-based technologies applied to mathematics education, relatively few recent studies exist (Shen et al., 2021a; Suresh et al., 2022; Loginova and Benoit, 2022) which explore transfer learning to improve language representation for math-related tasks (Peng et al., 2021; Shen et al., 2021b). Among these, MathBERT (Shen et al., 2021b) has been built specifically for challenging downstream NLP tasks in math education (e.g., knowledge component prediction, auto-grading open-ended question-answering, and knowledge tracing). It is indeed a mathematics-customized BERT model (Devlin et al., 2019). MathBERT representations are created by pre-training the BERT-base model on large mathematical corpora, including pre-kindergarten, to high-school and college graduate-level mathematical text.

2.2 Dialogue Systems and NLU

For interactive early math learning applications, as we aim to build a spoken dialogue system for kids, we will briefly discuss the existing dialogue system technologies and language understanding approaches in a more generic context here.

Conversational agents or dialogue systems are mainly categorized as either open-ended or task-oriented (Chen et al., 2017). The open-ended dialogue systems or chatbots allow generic conversations such as chit-chat (Serban et al., 2016; Jurafsky and Martin, 2018). On the other hand, task-oriented conversational AI systems are designed to accomplish specific tasks and handle goal-oriented conversations (Serban et al., 2018; Mehri et al., 2020). With the advances of deep learning-based language technologies, improved access to high computing power, and increased availability of large datasets; the end-to-end trained dialogue systems can achieve encouraging results for both open-ended (Serban et al., 2016; Dodge et al., 2016) and task-oriented (Wen et al., 2017; Bordes et al., 2017; Ham et al., 2020) applications. Dialogue Managers (DM) of task-oriented conversational AI systems are mostly sequential decision-makers. At that step, learning the optimal dialogue policies can be achieved via reinforcement learning (RL) from an excessive number of user interactions (Zhao and Eskenazi, 2016; Shah et al., 2016; Cuayáhuitl, 2017; Dhingra et al., 2017; Liu et al., 2017; Su et al., 2017). However, building RL-based dialogue systems with highly limited user interaction data is immensely challenging. For this reason, supervised learning methods with traditional pipeline-based modular dialogue systems are still widely accepted when training data is initially limited to bootstrap the task-oriented SDS for further data collection (Budzianowski et al., 2018). For implicit dialogue context management, statistical and neural network-based dialogue system frameworks and toolkits (Bocklisch et al., 2017; Ultes et al., 2017; Burtsev et al., 2018) are employed popularly in the research communities and industrial applications.

The NLU module of a dialogue system pipeline processes the user utterances as input text and usually predicts the user's intents or dialogue acts. For sequence learning tasks of Intent Recognition and Slot Filling (Mesnil et al., 2015; Hakkani-Tür et al., 2016), joint training of Intent Classification and Entity Recognition models have been explored widely (Zhang and Wang, 2016; Liu and Lane, 2016; Goo et al., 2018; Varghese et al., 2020). Many hierarchical multi-task learning architectures have been proposed for these joint NLU methods (Zhou et al., 2016; Gu et al., 2017; Wen et al., 2018; Okur et al., 2019; Vanzo et al., 2019). Vaswani et al. (2017) proposed the Transformer as a game-changing neural network architecture based entirely on attention mechanisms (Bahdanau et al., 2015). Right after the Transformers, Bidirectional Encoder Representations from Transformers (BERT) (Devlin et al., 2019) has been proposed. BERT has become one of the most significant breakthroughs in language representations research and has shown strong performance in several NLP tasks, including NLU. Lately, Bunk et al. (2020) introduced the Dual Intent and Entity Transformer (DIET) model as a lightweight multi-task architecture. DIET has been shown to outperform fine-tuning the BERT model for predicting intents and entities on a complex multi-domain NLU-Benchmark dataset (Liu et al., 2021). For efficient language representation learning, Henderson et al. (2020) recently proposed the Conversational Representations from Transformers (ConveRT) model, which is another lightweight approach to obtain pre-trained embeddings as sentence representations successfully used in several conversational AI tasks.

3 Language Understanding Methods

For the NLU module within our early math-learning dialogue system pipeline, we have examined numerous options for Intent Recognition and built our NLU models on top of the Rasa open-source framework (Bocklisch et al., 2017).

3.1 Baseline: TF+BERT

The previous baseline Intent Recognition architecture available in the Rasa platform was based on supervised embeddings as part of the Rasa NLU (Bocklisch et al., 2017). It was an embedding-based text classifier that embedded user utterances and intent labels into the same vector space. This former baseline architecture was inspired by the StarSpace algorithm (Wu et al., 2018), where the supervised embeddings were trained by maximizing the similarity between intents and utterances. Sahay et al. (2019) enriched this embedding-based former baseline Rasa Intent Classifier by incorporating additional features and adapting alternative network architectures. To be more specific, they adapted the Transformer architecture (Vaswani et al., 2017) and employed pre-trained BERT embeddings (Devlin et al., 2019) using the `bert-base-uncased` [2] model for Intent Recognition. We treat this simple and initial approach as our baseline NLU model in this study. We will refer to this method as TF+BERT in our experiments.

3.2 DIET+BERT

Next, we explored the potential benefits of adopting the recent DIET architecture (Bunk et al., 2020) for the Intent Recognition task in the basic math-learning domain. DIET is a transformer-based multi-task architecture for joint Intent Classification and Entity Recognition. The architecture includes a two-layer transformer shared for both NLU tasks. A sequence of entity labels is predicted with a Conditional Random Field (CRF) (Lafferty et al., 2001) tagging layer on top of the transformer output sequence corresponding to the input sentences treated as a sequence of tokens. For the intent labels, the transformer output for the classification token and the intent labels are embedded into the same semantic vector space. The dot-product loss is employed to maximize the similarity with the target label and minimize similarities with the negative samples. DIET architecture enables the incorporation of the pre-trained word and sentence embeddings from language models as dense features, with the flexibility to combine these with token-level one-hot and multi-hot encodings of character n-grams as sparse features. These sparse features are passed through a fully-connected layer with shared weights across all sequence steps. The output of the fully-connected layer is concatenated with the dense features from the pre-trained models. The high flexibility of this architecture allowed us to use any pre-trained embeddings as dense features in DIET, such as BERT (Devlin et al., 2019), ConveRT (Henderson et al., 2020), and MathBERT (Shen et al., 2021b). To investigate the net benefits of DIET architecture, we adopted DIET with off-the-shelf pre-trained BERT embeddings using the `bert-base-uncased` model (Devlin et al., 2019) and compared that against our baseline TF+BERT model. This approach (i.e., combining out-of-the-box BERT representations with the DIET classifier) will be referred to as DIET+BERT in the experiments[3].

[2] `https://huggingface.co/bert-base-uncased`

[3] Refer to Bunk et al. (2020) for hyper-parameters, computational costs, and hardware specifications.

3.3 DIET+ConveRT

Conversational Representations from Transformers (ConveRT) (Henderson et al., 2020) is a promising architecture recently proposed to learn pre-trained representations that are well-suited for conversational AI applications, especially for the real-world Intent Classification tasks. ConveRT is a transformer-based dual-encoder network leveraging quantization and sub-word level parameterization, where the pre-trained representation outputs from its sentence encoder can be utilized for the conversational Intent Classification tasks. DIET and ConveRT are both lightweight architectures with faster and more efficient training capabilities than their counterparts. When incorporating the ConveRT embeddings within the DIET classifier, the initial embeddings for the classification tokens are set as the input sentence encoding obtained from the ConveRT model. That enables exploiting extra contextual information from the complete sentence on top of the word embeddings. For these reasons, we adopted the DIET architecture and utilized pre-trained ConveRT embeddings to potentially improve the Intent Recognition performances on our domain-specific early math-learning datasets. We will call this approach DIET+ConveRT in our experiments.

3.4 DIET+MathBERT

Finally, a pre-trained language model named MathBERT (Shen et al., 2021b) has been presented lately for downstream NLP tasks in the domain of mathematics education. MathBERT is a BERT-like language representation model further pre-trained from the `bert-base-uncased` model with dedicated mathematical corpora. Note that BERT is a general-purpose language model trained on a vast amount of unlabeled text corpus (i.e., Wikipedia and BookCorpus) with 3.3 billion words, which can be further pre-trained to obtain a new set of model weights for transfer learning. On the contrary, MathBERT is pre-trained on 100 million tokens of mathematical corpora, including instructional texts from books, curriculum, Massive Open Online Courses (MOOCs), and arXiv.org paper abstracts, covering all possible grade levels from pre-k to college graduate-level content. Although the scale of training corpora is much smaller than the BERT-base model, MathBERT can still have the potential to be more effective in math-related NLP tasks. That is hypothesized because mathe-

Statistics	Math-Game Data Planting	Watering
# Distinct Intents	14	13
Total # Samples (Utterances)	1927	2115
# Math-related Samples	452	599
Min # Samples per Intent	22	25
Max # Samples per Intent	555	601
Avg # Samples per Intent	137.6	162.7
Min # Words per Sample	1	1
Max # Words per Sample	74	65
Avg # Words per Sample	5.26	4.95
# Unique Words (Vocab)	1314	1267
Total # Words	10141	10469

Table 1: KidSpace-POC NLU Dataset Statistics

Statistics	Math-Game Data Planting	Watering
# Distinct Intents	12	11
Total # Samples (Utterances)	2173	2122
# Math-related Samples	549	602
Min # Samples per Intent	4	6
Max # Samples per Intent	1005	1005
Avg # Samples per Intent	181.1	192.9
Min # Words per Sample	1	1
Max # Words per Sample	45	44
Avg # Words per Sample	4.80	4.48
# Unique Words (Vocab)	772	743
Total # Words	10433	9508

Table 2: KidSpace-Deployment NLU Dataset Statistics

matical language frequently uses domain-specific vocabulary and concepts that require better word representations within the math context. With that motivation, we utilized the publicly available MathBERT release [4], which includes the PyTorch and TensorFlow versions of MathBERT models and the tokenizers. We explored adopting the DIET architecture with pre-trained MathBERT embeddings to empower the NLU/Intent Recognition task in our dialogue system designed for teaching basic math concepts. Furthermore, we investigated the representations from the MathBERT-base model that is trained with the BERT-base vocabulary (i.e., origVocab) and compared that against the representations from MathBERT-custom model pre-trained with math-specific vocabulary (i.e., mathVocab). This custom mathVocab set is also released to reflect the specific nature of mathematical jargon and concepts used in math corpora. In our NLU experiments, we will refer to these two distinct approaches as DIET+MathBERT-base (origVocab) and DIET+MathBERT-custom (mathVocab).

4 Experimental Results

4.1 Math-Game Datasets

Our experiments are conducted on the NLU datasets of Kid Space Planting and Watering use cases (Anderson et al., 2018; Aslan et al., 2022), having utterances from gamified math learning experiences designed for early childhood education (i.e., 5-to-8 years old kids). The intelligent conversational agent should accurately understand these children's utterances and provide appropriate feedback. The use cases include a specific flow of interactive games facilitating elementary math learning. The FlowerPot (i.e., Planting) game builds on the math concepts of tens and ones, with the larger flower pots representing tens and smaller pots representing ones. The virtual character provides the number of flowers the children should plant, and when the children have placed the correct number of large and small pots against the wall, digital flowers appear. In the NumberGrid (i.e., Watering) game, the children are presented with basic math questions with clues. When the correct number is touched on the 1-to-100 number grid projected on the wall, water is virtually poured to water the flowers. The virtual character supports the kids with learning to construct numbers using the 'tens' and 'ones' digits, practicing simple counting, addition, and subtraction operations. These math-game datasets have a limited number of user utterances, which are annotated manually for intent types defined for each learning activity. Some of the intents are highly generic across learning activities (e.g., *affirm*, *deny*, *next-step*, *out-of-scope*, *goodbye*), whereas others are highly domain-dependent and game-specific (e.g., *intro-meadow*, *answer-flowers*, *answer-water*, *answer-valid*, *answer-invalid*) or math-learning/task-specific (e.g., *ask-number*, *counting*).

The NLU models are trained and validated on the initial proof-of-concept (POC) datasets (Sahay et al., 2021) to bootstrap the agents for real-world deployments. These POC datasets were curated manually to train the initial SDS models based on User Experience (UX) studies. The models are then validated on the UX sessions in the lab with five kids going through these early math-learning games. Table 1 shows the statistics

[4]https://github.com/tbs17/MathBERT

Model	Math-Game Data Planting	Watering
TF+BERT (baseline)	90.50±0.25	92.43±0.32
DIET+BERT	94.00±0.38	96.39±0.14
DIET+ConveRT	**95.88±0.42**	**97.69±0.11**
DIET+MathBERT-base	90.40±0.16	93.56±0.26
DIET+MathBERT-custom	90.82±0.10	93.67±0.10

Table 3: NLU/Intent Recognition micro-avg F1-scores (%): TF+BERT (baseline), DIET+BERT, DIET+ConveRT, DIET+MathBERT-base (origVocab), and DIET+MathBERT-custom (mathVocab) models trained and validated on KidSpace-POC datasets.

Model	Math-Game Data Planting	Watering
TF+BERT (baseline)	85.08±0.49	90.06±0.56
DIET+BERT	87.03±0.30	89.63±0.62
DIET+ConveRT	**89.00±0.29**	**90.57±0.86**
DIET+MathBERT-base	84.52±1.19	86.85±1.15
DIET+MathBERT-custom	85.22±0.78	87.80±1.45

Table 4: NLU/Intent Recognition micro-avg F1-scores (%): TF+BERT (baseline), DIET+BERT, DIET+ConveRT, DIET+MathBERT-base (origVocab), DIET+MathBERT-custom (mathVocab) models trained on KidSpace-POC & tested on KidSpace-Deployment datasets.

of these KidSpace-POC NLU datasets. Planting and Watering game POC datasets have 1927 and 2115 user utterances, respectively. The deployment datasets were collected later from twelve kids, where the math-learning system was deployed in a classroom at school (Okur et al., 2022b). Table 2 shows the statistics of KidSpace-Deployment NLU datasets, where Planting and Watering deployment datasets have 2173 and 2122 user utterances, respectively. These deployment datasets are used only for testing purposes, where we train our NLU models on the POC datasets. For both in-the-lab and real-world datasets, the spoken user utterances and agent responses are transcribed manually at first. These transcriptions are annotated for the user intent and agent response types we defined for each math learning activity. These transcribed and annotated final utterances are analyzed and used in our experiments.

4.2 DIET Classifier with MathBERT Representations for NLU

Tables 3 and 4 present a summary of our NLU/Intent Recognition experimental results for the Planting and Watering math-game use cases at school, covering a series of task-oriented interactions for early math education. On the proof-of-concept (POC) datasets that we created for each math-game, we achieved above 95% F1-scores for Intent Recognition performances with our best-performing DIET+ConveRT NLU models (see Table 3). We trained and cross-validated these NLU models on math-game or activity-specific datasets having around 2K POC samples. When we later tested these models on real-world deployment data collected at school, we observed F1-score performance drops of around 7% with our best-performing DIET+ConveRT models (see Table 4). This performance drop is anticipated and explainable as we operate on the noisier real-world utterances collected from younger children in dynamic play-based environments.

To the best of our knowledge, this study presents the first attempt to adopt the lightweight multi-task DIET architecture and incorporate pre-trained MathBERT embeddings as dense features. We combine these MathBERT representations with sparse word and character-level n-gram features in a plug-and-play fashion. The motivation behind this was empowering math domain-specific embeddings for the NLU task targeted at kids playing basic math-learning games. However, we could not observe any benefits of employing MathBERT-base (i.e., pre-trained using origVocab of BERT-base) or MathBERT-custom (i.e., pre-trained using customized mathVocab) representations for the Intent Recognition task. Although MathBERT-custom seems to perform slightly better than MathBERT-base as expected, the gain is insignificant and still way lower than ConveRT and even BERT. There could be many reasons for this, such as the possible mismatch between our domain and advanced mathematical corpora (e.g., mathematical equations and symbols) with graduate-level textbooks that MathBERT trained on. Compared to this, our early childhood math education domain involves more basic concepts and simple operations (e.g., ones and tens, counting, adding, and subtracting). In addition, MathBERT is pre-trained on a relatively small set (i.e., 100M tokens) compared to the massive general-purpose corpora that the BERT models are trained on (3.3B words). ConveRT embeddings have already been shown to perform well on conversational tasks such as Intent Classification, partly because these are pre-trained on large cor-

NLU Model	Math-Game Activity	# Test Samples (Speech/Utterances)	NLU/Intent Recognition F1 (%)	ASR+NLU F1 (%)
DIET+ConveRT	Planting	588	91.8	75.8
DIET+ConveRT	Watering	664	97.4	84.7

Table 5: End-to-End (ASR + NLU/Intent Recognition) Evaluation Results on Planting and Watering Math-game activity datasets: DIET+ConveRT models trained on KidSpace-POC datasets and tested on KidSpace-Deployment datasets.

Math-Game Activity	#test-NLU (utterances)	NLU/Intent F1 (%)	ASR+NLU F1 (%)	#test-DM (responses)	DM/Response F1	NLU+DM F1	ASR+NLU+DM F1
Planting	184	90.5	73.3	209	0.89	0.87	0.82
Watering	346	95.2	84.4	403	0.93	0.91	0.89

Table 6: End-to-End (ASR + NLU/Intent Recognition + DM/Response Selection) Evaluation Results on Planting and Watering Math-game activity datasets: DIET+ConveRT NLU models and TED DM models trained on KidSpace-POC datasets and tested on KidSpace-Deployment datasets.

pora of natural conversational datasets (e.g., Reddit conversational threads). The success of ConveRT representations over MathBERT could also indicate that our educational game datasets involve numerous utterances around play-based conversations tailored towards planting and watering flower use cases. Compared to those, our datasets include limited interactions that directly involve numbers or counting/addition/subtraction operations. Our intent class distributions also support this observation, where we have around 450-600 samples within approximately 2K utterances in POC datasets annotated with directly math-related intents (e.g., *ask-number*, *counting*). In the deployment datasets, we observed around 550-600 math-related utterances within a total of 2.1K samples. Nevertheless, instead of pre-training the BERT models to create MathBERT, one can also explore pre-training the ConveRT model on large and more elementary math-related corpora as the next step.

4.3 End-to-End Evaluation

Our SDS pipeline starts by recognizing user speech via the ASR module and feeds the recognized text into our NLU component. We developed NLU models performing Intent Recognition to interpret user utterances. Then we pass these user intents together with multimodal inputs, such as user actions and objects, into the DM component. The multimodal dialogue manager handles verbal and non-verbal communication inputs from the NLU (e.g., intents) and external nodes processing audio-visual information (e.g., poses, gestures, objects, game events, and actions). We pass these multimodal inputs directly to the DM in the form of relevant multimodal intents. The Dialogue State Tracking (DST) model tracks what has happened (i.e., the dialogue state) within the DM. Then, the output of DST is used by the Dialogue Policy to decide which action the system should take next. Our DM models predict the appropriate agent actions and responses based on all the available contextual information (i.e., language-audio-visual inputs, game events, and dialogue history/context from previous turns). When the DM predicts verbal response types, the NLG module retrieves actual bot responses that are template-based in our use cases. We create a variety of response text by preparing multiple templates for each response type, where the final response template is randomly assigned at run-time. Finally, the generated text responses are sent to the TTS module to output agent utterances. Please refer to Okur et al. (2022a) for our multimodal SDS pipeline diagram.

We were assuming perfect (or human-level) speech recognition performances for the NLU results obtained on manual transcriptions (by human transcribers) until now. However, we observed around 30% word-error-rate (WER) in ASR transcriptions for kids' speech. These ASR outputs are obtained via the top hypothesis given by Google Cloud Speech-to-Text API [5]. Considering these ASR errors propagating into the ASR+NLU pipeline, the Intent Recognition F1-score performances drop around 11-to-17% (see Tables 5 and 6) when evaluated directly on noisy ASR outputs acquired using Google ASR engine. We are cur-

[5]https://cloud.google.com/speech-to-text/

Math-Game	Human Transcription	ASR Output	Intent	Prediction
Planting	Sunflowers	7 flour	*answer-flowers*	*counting*
	We need just one more	jasmine ranvir	*counting*	*answer-flowers*
	Twenty four	trailer for	*counting*	*intro-game*
	We just counted nineteen	he doesn't canton my team	*counting*	*out-of-scope*
	Twenty two	tell me too	*counting*	*out-of-scope*
	Six	snakes	*counting*	*answer-valid*
	Twelve	towel	*counting*	*answer-invalid*
	Nine	nah	*counting*	*deny*
Watering	Water could help them bloom	why don't can count them you	*answer-water*	*counting*
	Thirteen flowers	sure thing flowers	*counting*	*affirm*
	Seven	I haven't	*counting*	*deny*
	Twenty eight	try it	*counting*	*out-of-scope*
	Three	tree	*counting*	*answer-valid*
	Five	bye	*counting*	*goodbye*
	Bye bye Oscar	buy a hamster	*goodbye*	*answer-invalid*

Table 7: ASR + NLU/Intent Recognition Error Samples from Kid Space Planting and Watering Math-game datasets.

rently working towards improving the Automatic Speech + Intent Recognition (i.e., ASR+NLU) performances by exploring the N-best ASR hypotheses (Ganesan et al., 2021) instead of using only the top ASR hypothesis.

For the DM model development, we adopted a recently proposed Transformer Embedding Dialogue (TED) policy architecture (Vlasov et al., 2019), which is highly suitable to our multimodal math-learning use cases. In TED architecture, a transformer's self-attention mechanism operates over the sequence of dialogue turns to select the appropriate agent response. Despite the noisy ASR outputs with relatively higher WER in kids' speech compared to adults, when we performed end-to-end evaluations with the ASR+NLU+DM pipeline, we observed only 4-to-7% drops in response prediction F1-score performances (see Table 6). That means error propagation from ASR and NLU has much less effect on the Dialogue Manager (DM) outputs, which are the agent's final actions and selected responses.

4.4 Error Analysis

Table 7 presents several concrete examples to compare the utterance text obtained by manual human transcriptions (i.e., ground truth) versus problematic ASR outputs (i.e., speech transcriptions). The ground truth intent classes based on gold data annotations on human transcriptions are shown along with the predicted intent classes on ASR outputs obtained by our best-performing DIET+ConveRT NLU models. These ASR errors, especially on domain-specific math-related intents, could explain how errors propagate into the NLU module of our SDS pipeline and significantly degrade the performance of the Intent Recognition task. Although such ASR errors are expected in noisy real-world application data, especially with kids of age 5-to-8 (Dutta et al., 2022), this analysis also points to a significant room for improvements in the ASR+NLU pipeline. It also encourages us to explore mitigation strategies such as utilizing N-best ASR outputs (Ganesan et al., 2021) and employing phonetic-aware representations (Sundararaman et al., 2021) that can be more robust to ASR errors.

5 Conclusion and Future Work

Improving the quality of mathematics education is vital in developing problem-solving and critical thinking skills for younger learners, which are fundamental for comprehensive STEM education. This study showcased a task-oriented SDS built to promote play-based math learning for early childhood education. The conversational AI system is implemented and evaluated on real-world deployment data collected in classrooms while the kids are practicing basic math concepts. We presented our attempts to enrich the modular SDS pipeline for gamified math learning and prosecuted an end-to-end evaluation using several SDS components tested on real-world deployment data. For NLP applied to math education, language representations can play a significant role due to the exceptional nature of math language. We investigated employing the language representations from the MathBERT model created by pre-training the BERT-base on mathematical corpora. We compared the Intent Recognition results obtained using BERT and ConveRT representations versus the recently proposed

MathBERT embeddings on top of the DIET architecture for NLU. To perform an end-to-end evaluation of our SDS pipeline, we evaluated the ASR, NLU, and DM modules to investigate how error propagation affects the overall SDS performance in real-world math-learning scenarios.

In future work, we aim to explore adopting the N-Best-ASR-Transformer architecture (Ganesan et al., 2021) to utilize multiple ASR hypotheses. This approach can improve the Intent Recognition performances and mitigate recognition errors propagated into the ASR+NLU+DM pipeline due to using only the top ASR hypothesis. Another future direction is to enhance math-specific language representation learning by pre-training the ConveRT model on large math corpora, especially tailored towards early math education (e.g., pre-k to 2nd-grade math curriculum), and then fine-tuning them on our NLU tasks for game-based math learning.

Limitations

Before discussing the limitations of our study, note that the goal of this multimodal dialogue system that we built is to improve the quality of mathematics education for younger learners. To begin with, the cost of the overall setup currently deployed at school (e.g., projector, RGB-D/3D cameras, LiDAR sensor, lapel microphones) can be a limitation, especially for public schools in disadvantaged regions. That can potentially prevent us from having a broader positive impact with our AI for social-good efforts.

Another limitation of this work is the size of the collected datasets. Since the multimodal data collection from authentic classrooms and their labor-intensive annotation process is costly, we need to be innovative to work on such low-data regimes and benefit from the transfer learning paradigm whenever possible. Unfortunately, this data scarcity also affects the generalizability and reliability of our experimental results and end-to-end evaluations, which affects the overall robustness of such real-world applications.

In addition to the deployment costs and data-size concerns, we are bound to use lapel microphones to capture the speech from subjects. That affects the overall unobtrusiveness of the system. Although our ultimate goal is to use the microphone arrays in the classroom, the high WER observed in ASR with kids' speech, even with lapel mics, prevents us from using these far-field mic-array recordings.

Finally, instead of pre-training the BERT model on large math corpora (as performed to create MathBERT), we aimed to pre-train the ConveRT model on a more early-math-related subset of the corresponding corpora. However, although the code and models are publicly available, the authors of MathBERT (Shen et al., 2021b) have not released the fine-tuning math dataset per the data owner's request. Hence, we cannot perform these DIET+MathConveRT experiments before we collect our early-math-related corpora for transfer learning, which is another limitation of this work.

Ethics Statement

It is worth noting that before the proof-of-concept research deployments, a rigorous Privacy Impact Assessment process has been followed. Legal approvals have been sought and received to conduct research with parents, their children, and educators. All research participants signed consent forms before the studies, informing them of all vital details about the studies, including the research goals and procedures, along with how their data would be collected and used to support our research. As with all research projects, our collaborators abide by strict data privacy policies and adhere to ongoing oversight.

Acknowledgements

We gratefully acknowledge our current and former colleagues from the Intel Labs Kid Space team, especially Ankur Agrawal, Glen Anderson, Sinem Aslan, Benjamin Bair, Arturo Bringas Garcia, Rebecca Chierichetti, Hector Cordourier Maruri, Pete Denman, Lenitra Durham, David Gonzalez Aguirre, Sai Prasad, Giuseppe Raffa, Sangita Sharma, and John Sherry, for the conceptualization and the design of school use-cases to support this research. We also would like to show our gratitude to the Rasa team for the open-source framework and the community developers for their contributions that enabled us to conduct our research and evaluate proof-of-concept models for our usages.

References

Glen J. Anderson, Selvakumar Panneer, Meng Shi, Carl S. Marshall, Ankur Agrawal, Rebecca Chierichetti, Giuseppe Raffa, John Sherry, Daria Loi, and Lenitra Megail Durham. 2018. Kid space: Interactive learning in a smart environment. In *Proceedings of the Group Interaction Frontiers in Technology*,

GIFT'18, New York, NY, USA. Association for Computing Machinery.

Sinem Aslan, Ankur Agrawal, Nese Alyuz, Rebecca Chierichetti, Lenitra M Durham, Ramesh Manuvinakurike, Eda Okur, Saurav Sahay, Sangita Sharma, John Sherry, et al. 2022. Exploring kid space in the wild: a preliminary study of multimodal and immersive collaborative play-based learning experiences. *Educational Technology Research and Development*, pages 1–26.

Dzmitry Bahdanau, Kyunghyun Cho, and Yoshua Bengio. 2015. Neural machine translation by jointly learning to align and translate. In *International Conference on Learning Representations (ICLR 2015)*.

Ryan S Baker. 2021. Artificial intelligence in education: Bringing it all together. *OECD Digital Education Outlook 2021: Pushing the Frontiers with Artificial Intelligence, Blockchain and Robots*, pages 43–51.

Nathaniel Blanchard, Michael Brady, Andrew M. Olney, Marci Glaus, Xiaoyi Sun, Martin Nystrand, Borhan Samei, Sean Kelly, and Sidney D'Mello. 2015. A study of automatic speech recognition in noisy classroom environments for automated dialog analysis. In *Artificial Intelligence in Education*, pages 23–33, Cham. Springer International Publishing.

Tom Bocklisch, Joey Faulkner, Nick Pawlowski, and Alan Nichol. 2017. Rasa: Open source language understanding and dialogue management. In *Conversational AI Workshop, NIPS 2017*.

Antoine Bordes, Y-Lan Boureau, and Jason Weston. 2017. Learning end-to-end goal-oriented dialog. In *International Conference on Learning Representations (ICLR 2017)*.

Paweł Budzianowski, Tsung-Hsien Wen, Bo-Hsiang Tseng, Iñigo Casanueva, Stefan Ultes, Osman Ramadan, and Milica Gašić. 2018. MultiWOZ - a large-scale multi-domain Wizard-of-Oz dataset for task-oriented dialogue modelling. In *Proceedings of the 2018 Conference on Empirical Methods in Natural Language Processing*, pages 5016–5026, Brussels, Belgium. Association for Computational Linguistics.

Tanja Bunk, Daksh Varshneya, Vladimir Vlasov, and Alan Nichol. 2020. DIET: lightweight language understanding for dialogue systems. *CoRR*, abs/2004.09936.

Mikhail Burtsev, Alexander Seliverstov, Rafael Airapetyan, Mikhail Arkhipov, Dilyara Baymurzina, Nickolay Bushkov, Olga Gureenkova, Taras Khakhulin, Yuri Kuratov, Denis Kuznetsov, Alexey Litinsky, Varvara Logacheva, Alexey Lymar, Valentin Malykh, Maxim Petrov, Vadim Polulyakh, Leonid Pugachev, Alexey Sorokin, Maria Vikhreva, and Marat Zaynutdinov. 2018. DeepPavlov: Open-source library for dialogue systems. In *Proceedings of ACL 2018, System Demonstrations*, pages 122–127, Melbourne, Australia. Association for Computational Linguistics.

Aoife Cahill, James H Fife, Brian Riordan, Avijit Vajpayee, and Dmytro Galochkin. 2020. Context-based automated scoring of complex mathematical responses. In *Proceedings of the Fifteenth Workshop on Innovative Use of NLP for Building Educational Applications*, pages 186–192, Seattle, WA, USA → Online. Association for Computational Linguistics.

Ying-Hong Chan, Ho-Lam Chung, and Yao-Chung Fan. 2021. Improving controllability of educational question generation by keyword provision. *CoRR*, abs/2112.01012.

Maud Chassignol, Aleksandr Khoroshavin, Alexandra Klimova, and Anna Bilyatdinova. 2018. Artificial intelligence trends in education: a narrative overview. *Procedia Computer Science*, 136:16–24. 7th International Young Scientists Conference on Computational Science, YSC2018, 02-06 July2018, Heraklion, Greece.

Hongshen Chen, Xiaorui Liu, Dawei Yin, and Jiliang Tang. 2017. A survey on dialogue systems: Recent advances and new frontiers. *SIGKDD Explor. Newsl.*, 19(2):25–35.

Heriberto Cuayáhuitl. 2017. Simpleds: A simple deep reinforcement learning dialogue system. In *Dialogues with Social Robots*, pages 109–118. Springer.

Debajyoti Datta, Maria Phillips, Jennifer L. Chiu, Ginger S. Watson, James P. Bywater, Laura E. Barnes, and Donald E. Brown. 2020. Improving classification through weak supervision in context-specific conversational agent development for teacher education. *CoRR*, abs/2010.12710.

Kenny Davila and Richard Zanibbi. 2017. Layout and semantics: Combining representations for mathematical formula search. In *Proceedings of the 40th International ACM SIGIR Conference on Research and Development in Information Retrieval*, pages 1165–1168.

UN Desa et al. 2016. Transforming our world: The 2030 agenda for sustainable development.

Jacob Devlin, Ming-Wei Chang, Kenton Lee, and Kristina Toutanova. 2019. BERT: Pre-training of deep bidirectional transformers for language understanding. In *Proceedings of the 2019 Conference of the North American Chapter of the Association for Computational Linguistics: Human Language Technologies, Volume 1 (Long and Short Papers)*, pages 4171–4186, Minneapolis, Minnesota. Association for Computational Linguistics.

Bhuwan Dhingra, Lihong Li, Xiujun Li, Jianfeng Gao, Yun-Nung Chen, Faisal Ahmed, and Li Deng. 2017. Towards end-to-end reinforcement learning of dialogue agents for information access. In *Proceedings of the 55th Annual Meeting of the Association for Computational Linguistics (Volume 1: Long Papers)*, pages 484–495, Vancouver, Canada. Association for Computational Linguistics.

Sidney D'Mello and Art Graesser. 2013. Autotutor and affective autotutor: Learning by talking with cognitively and emotionally intelligent computers that talk back. *ACM Trans. Interact. Intell. Syst.*, 2(4).

Jesse Dodge, Andreea Gane, Xiang Zhang, Antoine Bordes, Sumit Chopra, Alexander H. Miller, Arthur Szlam, and Jason Weston. 2016. Evaluating prerequisite qualities for learning end-to-end dialog systems. In *International Conference on Learning Representations (ICLR 2016)*.

Satwik Dutta, Dwight Irvin, Jay Buzhardt, and John H.L. Hansen. 2022. Activity focused speech recognition of preschool children in early childhood classrooms. In *Proceedings of the 17th Workshop on Innovative Use of NLP for Building Educational Applications (BEA 2022)*, pages 92–100, Seattle, Washington. Association for Computational Linguistics.

Karthik Ganesan, Pakhi Bamdev, Jaivarsan B, Amresh Venugopal, and Abhinav Tushar. 2021. N-best ASR transformer: Enhancing SLU performance using multiple ASR hypotheses. In *Proceedings of the 59th Annual Meeting of the Association for Computational Linguistics and the 11th International Joint Conference on Natural Language Processing (Volume 2: Short Papers)*, pages 93–98, Online. Association for Computational Linguistics.

Chih-Wen Goo, Guang Gao, Yun-Kai Hsu, Chih-Li Huo, Tsung-Chieh Chen, Keng-Wei Hsu, and Yun-Nung Chen. 2018. Slot-gated modeling for joint slot filling and intent prediction. In *Proceedings of the 2018 Conference of the North American Chapter of the Association for Computational Linguistics: Human Language Technologies, Volume 2 (Short Papers)*, pages 753–757, New Orleans, Louisiana. Association for Computational Linguistics.

Arthur C Graesser, Shulan Lu, George Tanner Jackson, Heather Hite Mitchell, Mathew Ventura, Andrew Olney, and Max M Louwerse. 2004. Autotutor: A tutor with dialogue in natural language. *Behavior Research Methods, Instruments, & Computers*, 36(2):180–192.

Yue Gu, Xinyu Li, Shuhong Chen, Jianyu Zhang, and Ivan Marsic. 2017. Speech intention classification with multimodal deep learning. In *Canadian conference on artificial intelligence*, pages 260–271. Springer.

Dilek Hakkani-Tür, Gokhan Tur, Asli Celikyilmaz, Yun-Nung Chen, Jianfeng Gao, Li Deng, and Ye-Yi Wang. 2016. Multi-domain joint semantic frame parsing using bi-directional rnn-lstm. *Interspeech 2016*, pages 715–719.

Donghoon Ham, Jeong-Gwan Lee, Youngsoo Jang, and Kee-Eung Kim. 2020. End-to-end neural pipeline for goal-oriented dialogue systems using GPT-2. In *Proceedings of the 58th Annual Meeting of the Association for Computational Linguistics*, pages 583–592, Online. Association for Computational Linguistics.

Matthew Henderson, Iñigo Casanueva, Nikola Mrkšić, Pei-Hao Su, Tsung-Hsien Wen, and Ivan Vulić. 2020. ConveRT: Efficient and accurate conversational representations from transformers. In *Findings of the Association for Computational Linguistics: EMNLP 2020*, pages 2161–2174, Online. Association for Computational Linguistics.

Shifeng Huang, Jiawei Wang, Jiao Xu, Da Cao, and Ming Yang. 2021. Real2: An end-to-end memory-augmented solver for math word problems. In *Workshop on Math AI for Education (MATHAI4ED), 35th Conference on Neural Information Processing Systems (NeurIPS 2021)*.

Jiyou Jia, Yunfan He, and Huixiao Le. 2020. A multimodal human-computer interaction system and its application in smart learning environments. In *Blended Learning. Education in a Smart Learning Environment*, pages 3–14, Cham. Springer International Publishing.

Zhuoren Jiang, Liangcai Gao, Ke Yuan, Zheng Gao, Zhi Tang, and Xiaozhong Liu. 2018. Mathematics content understanding for cyberlearning via formula evolution map. In *Proceedings of the 27th ACM International Conference on Information and Knowledge Management*, pages 37–46.

Daniel Jurafsky and James H. Martin. 2018. *Ch 24: Dialog Systems and Chatbots. Speech and Language Processing: An Introduction to Natural Language Processing, Computational Linguistics, and Speech Recognition*, 3rd (draft) edition. Prentice Hall PTR, Upper Saddle River, NJ, USA.

Alice Kerry, Richard Ellis, and Susan Bull. 2009. Conversational agents in e-learning. In *Applications and Innovations in Intelligent Systems XVI*, pages 169–182, London. Springer London.

Aashish Kumar and Anoop Rajagopal. 2021. Phygital math learning with handwriting for kids. In *Workshop on Math AI for Education (MATHAI4ED), 35th Conference on Neural Information Processing Systems (NeurIPS 2021)*.

John D. Lafferty, Andrew McCallum, and Fernando C. N. Pereira. 2001. Conditional random fields: Probabilistic models for segmenting and labeling sequence data. In *International Conference on Machine Learning*, ICML, pages 282–289.

Sweta P Lende and MM Raghuwanshi. 2016. Question answering system on education acts using nlp techniques. In *2016 world conference on futuristic trends in research and innovation for social welfare (Startup Conclave)*, pages 1–6. IEEE.

James C. Lester, Eun Y. Ha, Seung Y. Lee, Bradford W. Mott, Jonathan P. Rowe, and Jennifer L. Sabourin. 2013. Serious games get smart: Intelligent game-based learning environments. *AI Magazine*, 34(4):31–45.

Bing Liu and Ian Lane. 2016. Attention-based recurrent neural network models for joint intent detection and slot filling. In *Interspeech 2016*, pages 685–689.

Bing Liu, Gokhan Tur, Dilek Hakkani-Tur, Pararth Shah, and Larry Heck. 2017. End-to-end optimization of task-oriented dialogue model with deep reinforcement learning. In *Conversational AI Workshop, NIPS 2017*.

Xingkun Liu, Arash Eshghi, Pawel Swietojanski, and Verena Rieser. 2021. *Benchmarking Natural Language Understanding Services for Building Conversational Agents*, pages 165–183. Springer Singapore, Singapore.

Ekaterina Loginova and Dries Benoit. 2022. Structural information in mathematical formulas for exercise difficulty prediction: a comparison of nlp representations. In *Proceedings of the 17th Workshop on Innovative Use of NLP for Building Educational Applications (BEA 2022)*, pages 101–106.

Behrooz Mansouri, Shaurya Rohatgi, Douglas W Oard, Jian Wu, C Lee Giles, and Richard Zanibbi. 2019. Tangent-cft: An embedding model for mathematical formulas. In *Proceedings of the 2019 ACM SIGIR international conference on theory of information retrieval*, pages 11–18.

Shikib Mehri, Mihail Eric, and Dilek Hakkani-Tür. 2020. Dialoglue: A natural language understanding benchmark for task-oriented dialogue. *CoRR*, abs/2009.13570.

Grégoire Mesnil, Yann Dauphin, Kaisheng Yao, Yoshua Bengio, Li Deng, Dilek Hakkani-Tur, Xiaodong He, Larry Heck, Gokhan Tur, Dong Yu, and Geoffrey Zweig. 2015. Using recurrent neural networks for slot filling in spoken language understanding. *IEEE/ACM Transactions on Audio, Speech, and Language Processing*, 23(3):530–539.

Eda Okur, Shachi H Kumar, Saurav Sahay, Asli Arslan Esme, and Lama Nachman. 2019. Natural language interactions in autonomous vehicles: Intent detection and slot filling from passenger utterances. *20th International Conference on Computational Linguistics and Intelligent Text Processing (CICLing 2019)*.

Eda Okur, Saurav Sahay, and Lama Nachman. 2022a. Data augmentation with paraphrase generation and entity extraction for multimodal dialogue system. In *Proceedings of the Thirteenth Language Resources and Evaluation Conference*, pages 4114–4125, Marseille, France. European Language Resources Association.

Eda Okur, Saurav Sahay, and Lama Nachman. 2022b. NLU for game-based learning in real: Initial evaluations. In *Proceedings of the 9th Workshop on Games and Natural Language Processing within the 13th Language Resources and Evaluation Conference*, pages 28–39, Marseille, France. European Language Resources Association.

Kulothunkan Palasundram, Nurfadhlina Mohd Sharef, Nurul Nasharuddin, Khairul Kasmiran, and Azreen Azman. 2019. Sequence to sequence model performance for education chatbot. *International Journal of Emerging Technologies in Learning (iJET)*, 14(24):56–68.

Shuai Peng, Ke Yuan, Liangcai Gao, and Zhi Tang. 2021. Mathbert: A pre-trained model for mathematical formula understanding. *CoRR*, abs/2105.00377.

Ana Cristina Pires, Fernando González Perilli, Ewelina Bakała, Bruno Fleisher, Gustavo Sansone, and Sebastián Marichal. 2019. Building blocks of mathematical learning: Virtual and tangible manipulatives lead to different strategies in number composition. *Frontiers in Education*, 4.

Srikrishna Raamadhurai, Ryan Baker, and Vikraman Poduval. 2019. Curio SmartChat : A system for natural language question answering for self-paced k-12 learning. In *Proceedings of the Fourteenth Workshop on Innovative Use of NLP for Building Educational Applications*, pages 336–342, Florence, Italy. Association for Computational Linguistics.

J. Elizabeth Richey, Jiayi Zhang, Rohini Das, Juan Miguel Andres-Bray, Richard Scruggs, Michael Mogessie, Ryan S. Baker, and Bruce M. McLaren. 2021. Gaming and confrustion explain learning advantages for a math digital learning game. In *Artificial Intelligence in Education*, pages 342–355, Cham. Springer International Publishing.

Saurav Sahay, Shachi H. Kumar, Eda Okur, Haroon Syed, and Lama Nachman. 2019. Modeling intent, dialog policies and response adaptation for goal-oriented interactions. In *Proceedings of the 23rd Workshop on the Semantics and Pragmatics of Dialogue*, London, United Kingdom. SEMDIAL.

Saurav Sahay, Eda Okur, Nagib Hakim, and Lama Nachman. 2021. Semi-supervised interactive intent labeling. In *Proceedings of the Second Workshop on Data Science with Human in the Loop: Language Advances, NAACL 2021*, pages 31–40, Online. Association for Computational Linguistics.

Iulian Serban, Alessandro Sordoni, Yoshua Bengio, Aaron Courville, and Joelle Pineau. 2016. Building end-to-end dialogue systems using generative hierarchical neural network models. In *Proceedings of the AAAI Conference on Artificial Intelligence*, volume 30.

Iulian Vlad Serban, Ryan Lowe, Peter Henderson, Laurent Charlin, and Joelle Pineau. 2018. A survey of available corpora for building data-driven dialogue systems: The journal version. *Dialogue & Discourse*, 9(1):1–49.

Pararth Shah, Dilek Hakkani-Tur, and Larry Heck. 2016. Interactive reinforcement learning for task-oriented dialogue management. In *Deep Learning for Action and Interaction Workshop, NIPS 2016*.

Jia Tracy Shen, Michiharu Yamashita, Ethan Prihar, Neil Heffernan, Xintao Wu, Sean McGrew, and Dongwon Lee. 2021a. Classifying math knowledge components via task-adaptive pre-trained bert. In *International Conference on Artificial Intelligence in Education*, pages 408–419. Springer.

Jia Tracy Shen, Michiharu Yamashita, Ethan Prihar, Neil T. Heffernan, Xintao Wu, and Dongwon Lee. 2021b. Mathbert: A pre-trained language model for general NLP tasks in mathematics education. *CoRR*, abs/2106.07340.

Kayleigh Skene, Christine M O'Farrelly, Elizabeth M Byrne, Natalie Kirby, Eloise C Stevens, and Paul G Ramchandani. 2022. Can guidance during play enhance children's learning and development in educational contexts? a systematic review and meta-analysis. *Child Development*.

Pei-Hao Su, Paweł Budzianowski, Stefan Ultes, Milica Gasic, and Steve Young. 2017. Sample-efficient actor-critic reinforcement learning with supervised data for dialogue management. In *Proceedings of the 18th Annual SIGdial Meeting on Discourse and Dialogue*, pages 147–157, Saarbrücken, Germany. Association for Computational Linguistics.

Yueqiu Sun, Tangible Play, Rohitkrishna Nambiar, and Vivek Vidyasagaran. 2021. Gamifying math education using object detection. In *Workshop on Math AI for Education (MATHAI4ED), 35th Conference on Neural Information Processing Systems (NeurIPS 2021)*.

Mukuntha Narayanan Sundararaman, Ayush Kumar, and Jithendra Vepa. 2021. Phoneme-bert: Joint language modelling of phoneme sequence and asr transcript. *arXiv preprint arXiv:2102.00804*.

Abhijit Suresh, Jennifer Jacobs, Margaret Perkoff, James H Martin, and Tamara Sumner. 2022. Fine-tuning transformers with additional context to classify discursive moves in mathematics classrooms. In *Proceedings of the 17th Workshop on Innovative Use of NLP for Building Educational Applications (BEA 2022)*, pages 71–81.

Kaveh Taghipour and Hwee Tou Ng. 2016. A neural approach to automated essay scoring. In *Proceedings of the 2016 Conference on Empirical Methods in Natural Language Processing*, pages 1882–1891, Austin, Texas. Association for Computational Linguistics.

Stefan Ultes, Lina M. Rojas Barahona, Pei-Hao Su, David Vandyke, Dongho Kim, Iñigo Casanueva, Paweł Budzianowski, Nikola Mrkšić, Tsung-Hsien Wen, Milica Gasic, and Steve Young. 2017. PyDial: A multi-domain statistical dialogue system toolkit. In *Proceedings of ACL 2017, System Demonstrations*, pages 73–78, Vancouver, Canada. Association for Computational Linguistics.

Andrea Vanzo, Emanuele Bastianelli, and Oliver Lemon. 2019. Hierarchical multi-task natural language understanding for cross-domain conversational AI: HERMIT NLU. In *Proceedings of the 20th Annual SIGdial Meeting on Discourse and Dialogue*, pages 254–263, Stockholm, Sweden. Association for Computational Linguistics.

Akson Sam Varghese, Saleha Sarang, Vipul Yadav, Bharat Karotra, and Niketa Gandhi. 2020. Bidirectional lstm joint model for intent classification and named entity recognition in natural language understanding. *International Journal of Hybrid Intelligent Systems*, 16(1):13–23.

Ashish Vaswani, Noam Shazeer, Niki Parmar, Jakob Uszkoreit, Llion Jones, Aidan N. Gomez, Lukasz Kaiser, and Illia Polosukhin. 2017. Attention is all you need. In *Advances in Neural Information Processing Systems 30: Annual Conference on Neural Information Processing Systems 2017, 4-9 December 2017, Long Beach, CA, USA*, pages 5998–6008.

Vladimir Vlasov, Johannes E. M. Mosig, and Alan Nichol. 2019. Dialogue transformers. *CoRR*, abs/1910.00486.

Liyun Wen, Xiaojie Wang, Zhenjiang Dong, and Hong Chen. 2018. Jointly modeling intent identification and slot filling with contextual and hierarchical information. In *Natural Language Processing and Chinese Computing*, pages 3–15, Cham. Springer International Publishing.

Tsung-Hsien Wen, David Vandyke, Nikola Mrkšić, Milica Gašić, Lina M. Rojas-Barahona, Pei-Hao Su, Stefan Ultes, and Steve Young. 2017. A network-based end-to-end trainable task-oriented dialogue system. In *Proceedings of the 15th Conference of the European Chapter of the Association for Computational Linguistics: Volume 1, Long Papers*, pages 438–449, Valencia, Spain. Association for Computational Linguistics.

Rainer Winkler, Sebastian Hobert, Antti Salovaara, Matthias Söllner, and Jan Marco Leimeister. 2020. *Sara, the Lecturer: Improving Learning in Online Education with a Scaffolding-Based Conversational Agent*, page 1–14. Association for Computing Machinery, New York, NY, USA.

Rainer Winkler and Matthias Söllner. 2018. Unleashing the potential of chatbots in education: A state-of-the-art analysis. In *Academy of Management Annual Meeting (AOM)*.

Ledell Wu, Adam Fisch, Sumit Chopra, Keith Adams, Antoine Bordes, and Jason Weston. 2018. Starspace: Embed all the things! In *Proceedings of the Thirty-Second AAAI Conference on Artificial Intelligence*, volume 32.

Ke Yuan, Dafang He, Zhuoren Jiang, Liangcai Gao, Zhi Tang, and C Lee Giles. 2020. Automatic generation of headlines for online math questions. In *Proceedings of the AAAI Conference on Artificial Intelligence*, volume 34, pages 9490–9497.

Xuesong Zhai, Xiaoyan Chu, Ching Sing Chai, Morris Siu Yung Jong, Andreja Istenic, Michael Spector, Jia-Bao Liu, Jing Yuan, and Yan Li. 2021. A review of artificial intelligence (AI) in education from 2010 to 2020. *Complexity*, 2021.

Xiaodong Zhang and Houfeng Wang. 2016. A joint model of intent determination and slot filling for spoken language understanding. In *Proceedings of the Twenty-Fifth International Joint Conference on Artificial Intelligence*, IJCAI'16, pages 2993–2999. AAAI Press.

Zheng Zhang, Ying Xu, Yanhao Wang, Tongshuang Wu, Bingsheng Yao, Daniel Ritchie, Mo Yu, Dakuo Wang, and Toby Jia-Jun Li. 2021. Building a storytelling conversational agent through parent-ai collaboration.

Tiancheng Zhao and Maxine Eskenazi. 2016. Towards end-to-end learning for dialog state tracking and management using deep reinforcement learning. In *Proceedings of the 17th Annual Meeting of the Special Interest Group on Discourse and Dialogue*, pages 1–10, Los Angeles. Association for Computational Linguistics.

Qianrong Zhou, Liyun Wen, Xiaojie Wang, Long Ma, and Yue Wang. 2016. A hierarchical lstm model for joint tasks. In *Chinese Computational Linguistics and Natural Language Processing Based on Naturally Annotated Big Data*, pages 324–335, Cham. Springer International Publishing.

Author Index